To Gordon

In appreciation of his
wonderful artistic talent
and thanks for the use of his
superb picture as a cover.

Joe
8/26/10

PARIS TO TOKYO ON A DOLLAR AND A PRAYER

HITCHHIKING IN 1955

PARIS TO TOKYO ON A DOLLAR AND A PRAYER

HITCHHIKING IN 1955

Written by Joe Di Bona

Edited by Eleanor J. Harrington-Austin

Righter Books

Righter Publishing Company, Inc.
Post Office Box 105
Timberlake, NC 27583

www.righterbooks.com

July 2010

Printed and bound in the
United States of America

Library of Congress Control Number
2010931327

ISBN: 978-1-934936-58-0
Paris To Tokyo On A Dollar And A Prayer
By Joe Di Bona

Oh, I got plenty of nuthin'
And nuthin's plenty for me.

From "Porgy and Bess"
By Ira and George Gershwin

To Helene Rosenberg Di Bona
Who always came through

CONTENTS

INTRODUCTION

This memoir, written in 2010, recounts 1955, when I traveled from France to Tokyo and then to San Francisco, looking for something that was never very clear, even in my own mind. I began writing this memoir as a record of what I did and where I went, but that was never the important part of what I was trying to do. The plain fact is that I wanted to relive the life I had lived when I was 20. This is a record of how we mold ourselves through contact with extraordinary people who have qualities that most people lack. There are individuals who sustain an aura, the mere contact with which can change us forever. The men and women of ineffable spiritual force whom I encountered back in 1955 changed me forever. While it is difficult for me, even now, to understand what happened when I was in my twenties, I know that I will never escape the impact those persons and events had on my life.

Leaving New York City for Paris, 1953

In this memoir, I offer places on a map and names of cities, some of which have changed over the years and others that I cannot recall. They serve only as markers in a travel log, as a general guide, and as a reflection of the trip. The names of people I met or towns I passed through are occasionally only approximates. As I began writing I was impressed with how much I was enjoying the memories, which just seemed to flood back to me in rapid sequence as I typed. I was reliving those days as fully as I did so long ago, but, this time, without the anxieties or concerns I felt back then.

As I wrote, I confided to a friend that at points I just could not remember what happened next. She said, "Well, why don't you look up your 1955 notes and read them." I had indeed at the time kept occasional notes of my travels, but they were fragmentary and I feared they might lead me away from the main ideas I wanted to bring out. So I was reluctant to open the old, tattered journals that contained my original notes. However, when I did begin to read them, I quickly found them useful. They even recalled some things that I had completely forgotten. One such example involved the Punjabi spinning songs, translated for me as we sat together with a group, spinning by candlelight, in Sevagram, India. Each day some of the older women sang the songs they recalled of their childhood. I had completely forgotten these happy moments of songs and tales, and I was happy to find the songs in my notes and include them in this book.

So, what came out and what you are now about to read is an amalgam of memory, notes, and a desire to find meaning in what happened so long ago. What surprises me, even now, is how these strange travels came about in the first place. Remember that before my journey I knew nothing about the places I was to visit. Nor did I have any goal in mind. Someone told me to go to India or to Japan, and that was all the encouragement I needed. After all, I had worked for the Ford Motor Company for a couple of years after college, and, even though I was not unhappy, on my own I had decided I wanted to find my own life. So with that rather unformed idea in mind, I was off to Paris to study Chinese and learn about whatever life offered. Back then I was sure that when I was in my 80s I would not be able to do what I could do in my 20s, or so I thought. But to tell you the truth, at 83 years of age, I am still looking for adventure, not exactly as I did at 23, but life is still as exhilarating as ever. I only mention this to encourage others of any age to do what they wish, ignoring seeming social, physical, or financial blocks. You will never regret breaking the bonds of convention and conformity, although doing so is not as easy today as it was then. I have often encouraged my students at Duke University, who are dissatisfied with the academic routine or have a failed romance or suffer family pressures or fret over future career choices, to break away and seek their own lives, wherever those lives may be. Invariably, they are fearful of consequences.

"But how will we live?" they ask. And, alas, they are disappointed when I cannot tell them. They want answers and I have none to give that they can understand. Nevertheless, I am hopeful that this book will provide a few readers with the magic word that will wrest them from the world that is not theirs and show them the possibility of personal liberation.

Student Identification Card

I cannot refrain from adding a word on how much generosity I encountered in 1955. Again and again, I was helped by people in Europe, from France to Greece to Serbia and everywhere else. They did not know me, but made wherever I went possible. Today I would not dare try to hitchhike--as I and my companion, Dorothy, did then in 1955. In that post WWII time, America was regarded as a wonderful champion of liberty and prosperity and justice and faith. Today, that is no longer the case. Everywhere Americans travel today; there are terrorists and other dangerous people who think nothing of harming visitors. When I have spoken to my students, urging them to go off on their own, they remind me of the daily kidnappings, rapes, and murders that plague the world of innocent travelers. The students' estimate of our contemporary world is more current than mine, but, instead of looking at the way America has lost its leadership, they cannot imagine a world that was essentially happier and more peaceful than today. And perhaps that is the message of this book: to remind us that the world has not always been as treacherous as today. And, indeed, if this book is a record of peace and harmony in the past, it may be possible to see hope for the future.

If the world has changed so radically in the past half century, I have to acknowledge my great fortune to have had such an experience. I often wonder if others will find the generosity I encountered. They well may ask, "How can people be so giving? It must be a fraudulent hope you are expressing. Aren't you making this all up?" They question the carefree unconcern that is so rare today. They even question the existence of altruism and their own readiness to trust their own goodness. What they know from the newspapers are the extravagant earnings some individuals in finance received in 2009. David Tepper got $4 billion (not million, but billion). John Paulson made $2.3 billion. The top 25 managers received a total of $25.3 billion. What are students to make of this? They do not think how society may benefit from this bonanza. They think, instead, "How can I get mine?" That greed was completely alien to the heroes of this memoir and, I daresay, to most sane men and women around the world.

Finding long-lost relatives in Sicily

So, what is the essence of what I am trying to say? I stress the spiritual life that I found in India and Japan. I argue that we can never be satisfied with the mere accumulation of money or goods. There is a realm of reality that pervades all lives, past and present and future, that will not only endure but also will bring us great satisfaction, joy, and fulfillment. In this book, you will meet some of these saintly manifestations that you may have thought are no longer even possible to imagine. Here they are for you. I am not a particularly religious person and have never had any use for organized religion, but what I found in Sivananda; Ramakrishna; The Mother in the Aurobindo ashram in Pondicherry; Gandhi's Sevagram; Bhave's selfless work; or my personal mentor, the Reverand Ogata-san in Kyoto, is impossible to define. All these together have made a formidable and lasting impact on my own mental and spiritual make up. After all these years, I still see and hear the words they whispered into my ear again and again. So maybe that is what I am getting at. If I have failed in my rationale for bringing out a memoir such as this, perhaps you will tell me if I am on the mark or not.

I wish to thank the many people who have helped and guided my efforts with this book. I especially thank Eleanor J. Harrington-Austin, for editing this book; Rina Hutchinson, for proofing it with great care; Barbara Alston, for formatting the text, pictures and cover; and my friends of the Chapel Hill Writers' Group, for their encouragement and chapter-by-chapter reviews of my work.

GENOA, ITALY, RETURNING TO PARIS

The crowded Genoa youth hostel was filled with foreigners, sitting at small tables, telling one another where they had been or where they were headed next. The air was full of tobacco smoke, and the chairs and tables so close that it was hard to squeeze by to find an empty seat. With luck, Joe found a place, but, in approaching it, he bumped the young woman who was already seated and apologized as well as he could.

"It's all right," she responded, matter-of-factly, closing her notebook and turning her head to see what awkward character had come along. "By any chance, are you a writer?" she asked unexpectedly.

"No, I'm afraid not, although I guess you could call me a wannabe." Joe felt pretty comfortable in the hostel, having spent the last couple of months traveling up and down the west coast of Italy and then on to Sicily, where he visited some long lost relatives. So he did not mind telling this total stranger one of the most intimate secrets of his life.

"Well, I guess that will have to do. Sit down, but first see if you can find some more wine for me before we share our great ideas with one another."

By the time Joe returned with a fresh glass of wine, she seemed to have prepared a list of questions that began with "Where are you from? What do you do for a living? Do you really write or just play at it? And do you believe in the stars or read your horoscope?"

"Well, you certainly have a unique way to begin a conversation," Joe said. "How about asking me my name first?"

"Why would I want to do something so trivial and unimportant? If that's all that's on your narrow mind, maybe we should just end this never-going-anywhere relationship before we waste any more time being nice to one another."

Joe sat down without responding to the young woman and couldn't figure out whether she was being serious or just having fun with him. In either case, it didn't matter, since he was going to enjoy his glass of wine and then go up to sleep. But before doing that, he thought he could enjoy more of her unorthodox approach to socializing. He had read something that Carl Jung wrote about how you learn more about yourself by confronting people who are very different from you than from always remaining with people who are like clones of your mental makeup. He quoted an instance in which he had spoken to a Native American Indian in New Mexico who shook him with new ideas. The man maintained that Europeans were ugly-- with the wrong color, sharp noses, and thin lips-- and could never be content with being who they were.

He noticed the young woman had closed her notebook, so he asked her what she was writing about.

"It's none of your business," she answered, "and why do you want to know anyway?"

"Well, I guess I'm curious, especially so since you have your own way of dealing with strangers and I'm intrigued by your unique charm."

Chuang Tze dreaming he was a butterfly, or a butterfly dreaming he was Chuang Tze.

"Oh, so now I'm charming, am I? A moment ago you were looking at me as if I belonged in a museum." She was not smiling now and seemed to be getting ready for battle. Joe was not at all put off by her belligerence and thought to himself, "Let's try again." But, instead he said, "So what can you tell me about your writing?"

"Nothing, really, that you would understand, such as Chuang Tze..."

Joe was suddenly galvanized, "Did you say Chuang Tze?" He was practically half out of his chair and gasped, "The Chinese philosopher who dreamt he was a butterfly?"

Now it was the woman who was surprised. "I've been looking everywhere for months, hoping to find someone who I could talk to about Chuang Tze, but never would have thought that you would be the man I sought. Can we get out of here and find a quieter place where we can talk?"

And they did--until three in the morning, when they both just lay down on the outside benches and fell fast asleep. They awakened the next morning with the cleaning man watering the plants and sweeping up the debris from the night before. They were still exhausted, so they took a room in a cheap hotel and resumed their much-needed sleep. When they awoke about noon, each of them was similarly disoriented. Joe wondered where he was, how he got there, and who this strange woman beside him might be. The young woman had similar thoughts. She asked herself how she had gotten into this hotel, why they both were completely clothed, and who exactly this man next to her was.

As they both stirred, Joe turned to face his companion and said, "I'm Joe, and who do I have the pleasure of waking up facing?"

"I'm Dorothy, I'm English, and I am currently without goals or purpose in my life. So what is your story?"

"Not much different. I have neither goals nor purpose in my life, nor do I care. Things you want or strive for are a waste of time and take energy away from what is important."

"Like what?" Dorothy wanted to know.

"Like watching clouds and dreaming of the lives of birds or flowers, visiting museums to learn about the history of mankind, or meeting artists or listening to music or writing poems and so much more."

Dorothy spoke, "Yes, that's what I want to do, but my parents said I was a dreamer and would never amount to anything, so they gave me thirty-three pounds and told me to get lost and never come home again. So, tell me how you happen to do all those wonderful things without working. Are you a thief or something?"

"Not really. I do work, and, if you are serious about a new life, I'll tell you more about how I buy books at Paris auctions, and I mean books by the laundry cart, take them home, and resell them to specialized dealers. That's a start, but I still have some cash left from my job in New Jersey, enough to get me to India next."

"To India!" Dorothy was aghast. "I've always dreamt of going there, but it will take more money than I have or expect to have."

"You've got to start thinking outside the box. First of all, I will just hitchhike there. Look at a map and see there are roads directly from Europe to there, so that's not a problem. As far as sleeping goes, you sleep wherever you can. That's never a problem either."

"Where do you get these screwy ideas anyway? They sound okay but I never met anyone who believed in such notions."

"Well, I have to admit, I have been influenced by a great Hungarian mystic I met in Paris, named Anton Biro. He is an artist and lives at La Ruche with some pretty famous painters, including my heartthrob, Lea. Anyway, when I told him I was worried about soon not having enough money to live on, he told me to go where everyone is poor, as in India, and I will be perfectly happy living a much simpler and healthier life. So as soon as I get back to Paris, I'm picking up my stuff and hitting the road." As he ended these words, he turned toward Dorothy and began to unbutton her jacket.

"What are you doing?" Dorothy demanded in alarm.

"I'm removing your clothes, which is often preparatory to having sex."

"Oh, you may think so, but not with me you're not. I don't know you and I'm still not sure where this is headed, so let's give this some time, okay?"

"Okay with me. Let's get out of here and get some coffee and a brioche."

PARIS

So, as a cost-cutting measure, Dorothy and Joe returned to Paris, and she moved into his sort of boarding house where Mme. Barnes rented rooms on the second floor. Mme. Barnes lived downstairs, where she could keep an eye on her pantry, fearful that some hungry tenants might help themselves to her larder. Upstairs were Mme. Geysi, a Jewess who had lived through the Nazi era in Vichy, France, and barely got out of Germany to save herself. She had been the mistress of one of the Third Reich historians and, when she was discovered, she left and never found out what became of Helmut, her lover. The second resident was Herta, a talented painter who sold nothing. She went to flea markets, where her astute knowledge of ceramics enabled her to earn a living reselling sometimes valuable pieces to tourists. And, of course, Joe was the third resident, who had just enlarged his room by breaking a wall and was trying to get rid of the plaster and debris before anyone discovered what he had done. But Mme. Barnes traced the white powder from plaster left on shoes, and when Joe came down one morning, she said she had something to discuss with him. She was weighing a potato to be sure she did not consume more than her small frame required.

"Monsieur Josef, is he well today?" she asked, using the third person since the *tu* was out of the question and even the *vous* form too intimate. She continued, in a business-like manner, to say she was raising his rent 10 francs per month and more if he required an added electric outlet. That was fine with Joe, since his book dealing was going well and Dorothy was helping to sort out some of the books.

Villa Collette, Paris 1955

Then, without warning, Lea was back from Israel, where she had attended the reunion of her military unit that she had fought with just when Israel was most vulnerable.

She came directly to the Rue Villa Collette and marched into Joe's now two-room suite where Dorothy and Jack were on the floor surrounded with piles of books and putting flower picture books in one pile and nineteenth-century medical books in another. As soon as she walked in the door, Lea spoke to greet both Dorothy and Joe.

"So this is the slut who thinks she is going to take my place." Joe rose to Lea, but Dorothy had no idea what she was up to.

"Lea, Lea," he said, "this is not at all what you are thinking, so calm down and you can meet my assistant, Dorothy."

Lea and Anton Biro

"So that's what you call a live-in woman in your apartment. An Assistant? What do you take me for, nincompoop? I don't care what you call her; I want her out of here and fast. Do you hear me?" Now Lea was standing over Dorothy, who looked terribly frightened and was moving slowly away from the pile of books and into a corner.

"Look, Lea," Joe spoke in his quiet and sincere voice, "I can explain everything, but just give me a chance. I haven't seen you for weeks, so let's begin with a hug and a kiss." Somewhat reluctantly Lea accepted, and they embraced

31

warmly. "Please don't call Dorothy names. She is helping me and needs a friend in Paris. You know I love you and am happy you are back. We'll have dinner tonight and you can spend the night here. Is that O.K.?"

"So what do you do with your assistant? While we are making love, what is she doing? Just watching?"

"Oh, Lea, she can sleep in Herta's place next door."

Dorothy spoke up for the first time, "Oh, that's fine with me. I won't mind at all."

"We'll see about that," Lea answered. "But first I'd like a word with Dorothy myself, so why don't you take a walk, Joe, and give us a chance to get to know one another."

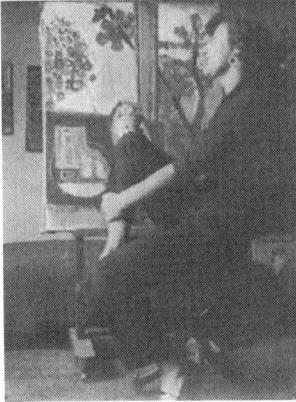

Lea

As soon as Joe was out of the house, Lea spoke directly to Dorothy. "So Joe says you're his assistant and is sort of saying there is no sex between the two of you, is that right?"

"Yes, that's true. We eat together, we work together, and I live here to save money. Joe has been a big help to me, and I don't know what I would have done without him?"

"You know, when you say it, I believe you, but, when Joe says it, I have my doubts. This is

32

not the first time he has embellished the truth for his own satisfaction. Has he ever tried to have sex with you?"

"Yes, when we first met. It was in Italy and we shared a room where he tried removing my clothes, but I said no, I didn't know him, and didn't think that we should start like that, and ever since he has often spoken of you and calls you his 'heartthrob.'"

"Well, it sounds okay, so I guess you can stay here, so long as you are not around when I am spending the night. Is that clear?"

"Certainly," Dorothy said with some comfort, "and thanks for your understanding."

After a few weeks, a certain routine emerged between the book buying and Lea's art and Dorothy's waiting to start on the fabled trip to India.

"I want to go as soon as you are ready," she said to Joe as he prepared their morning coffee with too much sugar and as much whiskey. Dorothy wondered why he always kept two small bottles of whiskey in the kitchen cabinet above the sink. She didn't want to ask him, but she noticed that in the morning he poured a drop from one bottle and in the evening he drank from the second bottle; and never confused the two.

"I do too," Joe answered. "Our money is drying up fast and the book business is nothing to write home about. Even the tutoring job you have is barely enough to pay for food."

"So, what is holding us up?" Dorothy wanted to know.

"I don't really know," Joe answered honestly. "Maybe I just need a kick in the pants to get moving, or a word from Biro or Lea. I tell you what, let's meet tonight at the Dome and tell everyone we're leaving tomorrow. I'll tell Lea, Mme. Geysi, Herta, Biro, and anyone else who cares to know."

Dorothy was taken aback at the suddenness of the idea and couldn't quite understand that Joe was really serious. But since he didn't really know himself, it couldn't be that important.

That evening, by some strange telepathy, the whole crew was present. Lea sat next to Joe and whispered softly in his ear that she missed him terribly and swore that if he really went to India she would kill herself. Biro said going was a great idea.

Party before leaving Paris

"Get out of Paris," he urged. "It is a hell hole and you see thousands of people who lose themselves in this supposed great cultural salvation. They need to go home and take up a normal life, which is all they are capable of." Biro was wearing socks and shoes and felt comfortable, despite the fact that he was the only one who believed that India could save your soul. Dorothy spoke up, asking why India was so different from, say, Japan or Tibet or Bulgaria.

Saying goodbye to friends in Paris

"India," Biro insisted, "is where the world began. It's all written down and translated in the Rig Vedas, which have come down to us unchanged as an oral transmission for five thousand years. So if anyone wants to redeem himself, he must first go to India and then learn a few sounds so the world will continue without end."

Lea wasn't at all convinced and spoke out, "Why not let the world end?" she asked. "It's in such a mess already, it can't get any worse. In the war the Arabs only wanted to kill us and we wanted to do the same to them. Is that what you want to save?"

Mme Geysi was the eldest in the group and said she had actually suffered the Holocaust and escaped with nothing but the clothes on her back. "But you," she said to no one in particular, "you have a chance to redeem the race of wrongdoers, so let them try. I am not saying they will save the world, but they may save themselves." And with those words, they all lifted their glasses in agreement and support, and, to underline that resolve, Herta took out her pocketbook and threw 1000 francs at Joe and Dorothy to be sure they would be on their way tomorrow. And the gesture was repeated by everyone else even by some unknown tourists who had been listening to their conversation from surrounding tables.

Leaving Villa Collette

Without realizing it, they were almost on their way. Lea insisted that Joe come home with her and Dorothy could pack her few belongings and be ready when Joe returned to Villa Collette in the morning to pick her up. And that was that.

FROM PARIS TO BASRA: EN ROUTE TO BOMBAY

In the morning, Joe and Dorothy took the metro to the bus station and were soon on the highway to Germany. Rides were plentiful until they came to Yugoslavia, where the war's destruction was still evident. The road was good, but there were no cars, busses, or trucks. None whatsoever. So they moved their bags to the middle of the road and sat there waiting. After an hour a bus came along and stopped. Since they blocked the road, there was nothing else the driver could do. When they sat down, he told them he was going to Skopie on the road to Greece, but they did not need tickets. This was a free ride for the defeat of Hitler by America in the war. Another passenger handed them fruit and some bread, and they exchanged the few words of English other passengers knew. Tourists were rare and hitchhikers practically unknown, so they were treated more like celebrities than poor students seeing the world.

Skopie turned out to be a more confusing town than they had anticipated. Someone on the bus had urged them to see Suli Han, described as a caravanserai near the baths and with a beautiful view of the Vardar River. When they found the place, it was largely in shambles from the war and in need of repairs. As they looked dolefully at what they had expected to be a lovely youth hostel, a man approached from one of the arched doorways and addressed them, first in Turkish and then, when he realized they did not respond, in English.

"I am Ahmed Shah and my family is responsible for this cultural treasure. Unfortunately, although we have asked for funds from the United Nations to rebuild Suli Nan, so far nothing is happening. Are you from England or perhaps Australia?"

"I am from the United States and Dorothy here is from England. We are hitch- hiking to India and this is where we have landed so far."

"I know something about different parts of the United States because during the war American cigarettes were in great demand among the German forces stationed in Skopie. My job was to buy them in Sofia and sell them here. It was a good job, and with the money I put away I have been able to afford three wives. I know Americans do not approve of multiple wives and, to tell you the truth, I am also against the practice now, but before I just followed the family custom."

"I'd like to know why you changed your mind." Dorothy asked.

"I will tell you," Ahmed responded. "But, first, come into the place to see the historic paintings. It has only two apartments that I have fixed up for tourists, but, since none have come yet, they remain vacant. In fact, if you need a place for tonight, I will be happy to let you stay in one without charge. I often use it when I want to get away from my own place with too many wives and noisy children."

"That is very generous of you." Joe said, "and I hope you get the money to fix up this treasure"

Ahmed led them through a large ornate hallway, but they could see where renovations were called for. The plaster was falling off the wall and accumulating on the floor, windows were broken, some of the supporting beams leaned dangerously and could crash down, and it was easy to see why it was far from ready to receive guests. The three walked around a corner where there was evidence that someone was looking after the place.

"This is Mumtaz," Ahmed said, pointing to a middle-aged woman wearing a scarf over her head and an apron and carrying a broom. "She helps keep the place clean and I have promised to pay her if we ever get some real money to fix the place up."

They entered one of the apartments and sat down at a small table. Without being summoned, Mumtaz brought coffee and some cookies, and, when they had relaxed, Dorothy again asked why Ahmed was no longer in favor of having more than one wife.

"First of all," he said, "none of my wives has any education, so there is never a question of having an intellectual discussion. They can discuss children and food and gossip about the neighbors, but political, economic, or cultural matters, about these things they are quite ignorant. Then, of course, there is a constant financial drain. Wives are costly to keep, you see." And here he looked directly at Dorothy. "You can talk to your husband and, if you want to work, you can do that too. It is wonderful and that is largely the reason I have changed my mind."

"I can understand that," Dorothy said, "but Joe and I are not married and I am not sure we ever will marry. For the moment, we enjoy each other's company and both of us are headed for India, and I would say we are happy. Wouldn't you agree, Joe?"

"I guess I would agree," Joe answered, but without the conviction that Dorothy used.

Ahmed could have explored the different responses but, instead, asked where they were headed next. Joe explained that often where they went depended on where their rides took them.

They really had no schedule and never knew where they would spend the night.

"Do you worry about where you are headed? Are you ever afraid, not knowing where you are going?"

"Oh, no," Joe answered quickly. "We think of this trip as an adventure, and we meet all sorts of fascinating people, and, if I may say so, like you. People we would never know if we worked in a school or a shop or anywhere else. And on top of that, we find most people are very helpful and generous. Without that innate altruism, we would still be in Paris, worrying about how much money we needed to live on."

Ahmed chuckled at what Joe had said and wondered how long such optimism would last, but for the moment and for their youth it seemed admirable.

"So where do you go next," he asked.

"Well, the road goes south and west, so it looks like Greece, Turkey, and then Aleppo."

"Ah, now, that's the place for me," Ahmed joined in, smiling. "It's beautiful. A real polyglot society bringing Armenians, Muslims, Jews, Turks, and everyone else under one roof, and it's been that way for thousands of years, and today it is the envy of anyone who visits even for a day. If you leave tomorrow, you should be in Turkey in a day or two, but don't count on it. You know how the Turks and Greeks hate each other, so much that they have never built a decent road between them, so you may end up taking a burro ride to cross the border."

"Is it really that bad?" Dorothy asked, almost to herself.

"Don't worry," Joe insisted. "We've been over some rough patches already and this won't be any worse."

The next morning, Ahmed took them out to a nice restaurant. He fed them well and arranged for a taxi to get them on the highway where they quickly found a ride to Thessalonica and the road to Turkey. But just as they got down from the car that had brought them and asked for the road to Turkey, suddenly everyone seemed to know nothing of any road that went there. They asked at shops, they knocked on doors, but no one seemed able to tell them which way to go. Finally, as they sat in a small café having a cup of tea, some students came by and sat down with them.

"Why do you want to go to Turkey?" one of them asked, and Joe told about going to India, hitchhiking all the way.

"That is almost impossible," another student insisted. "And, besides, when you go through Afghanistan there is fighting and it is a lawless area where they will steal your clothes and leave you to die of exposure."

"You look like nice people, but where did you ever get such ideas?"

"We've heard such stories even before we started in Paris and look where we are now: in

Greece and on the door step of Turkey and the Middle East. If you can't help us, perhaps you can tell us where we can find a bus to cross the border."

"We've told you that you are making a mistake, but since you are undeterred, we will introduce you to someone who knows how to get from here to there. You will have to leave at night, and, if the customs people stop him, he will claim all the contraband he carries belongs to you. Do you understand?"

"I don't like it, Joe," Dorothy insisted. "It's dangerous and sounds illegal. Why not wait till morning and see what else is available."

"Now, there is a sensible woman," one of the students agreed.

"Oh, let them go," one of the students called out. "Menelaus has made the trip often without incident, and he will be happy to have company. Talk to him and see what he says. We will wait here for your return. In the meantime, how about the shish kebob on us?"

When Menelaus did show up, it was already dark, and he stopped his car to get out and shake hands with everyone and then was introduced to Dorothy and Joe.

Looking at the empty dishes on the table, he said, "I'm glad you ate. We have a five-hour trip ahead of us, and we don't stop for anything on the way. Not that there is anywhere to stop. We take a little used route that is not monitored by either

Greek or Turkish officials. So if we are all ready, let's get on board."

The trunk was full, so Dorothy sat in back with their back packs filling most of the space, and Joe was in front with Mini, the name everyone called him. "Just remember," Mini warned, "though unlikely, we may be stopped by police, customs, or military. This is a route for illicit entry between Turkey and Greece, and you never know when or where things will happen. If it does, I will pretend to be a taxi service and all the parcels in the car belong to you. Agreed?"

The trip over the rugged mountain road took longer than a bus might take along the paved main road, but, once in Turkey, they stopped at a small café where the large parcels were unloaded, and Mini drove directly to the train station, where they had to wait for the next train in the morning. They slept fitfully in the station and entered the local train for the rest of the way to Ankara. It was full of workers going in to their jobs and, although Joe wanted to see about a ticket to the capital, there was no one selling tickets at the window, so they just entered with the rest of the crowd and stood holding onto straps since all the seats were filled. After an hour, more passengers came onto the train, and it grew so crowded that the riders could barely move.

One of the men in the crowd came up to Joe and demanded money. He spoke only Turkish and looked like every other worker in the train. Joe just turned away from him, but he came up to him a second time and demanded a couple of dollars, gesturing with his hands and looking very hostile. Joe thought this was the oldest trick in the book. You find some poor vulnerable foreigner and ask for money, and, if he doesn't pay, you threaten him as best you can. At the next stop, the same guy who was demanding money, got down and returned with two police who took Joe by his arms and carried him off the train.

"Where are you going, Joe?" Dorothy screamed.

"I have no idea, but I'll see you at the Youth Hostel in Ankara as soon as this is worked out. Just remember to call the American embassy and tell them what happened. I'll try to pay and get the next train, if I can."

That turned out not to be an option. Joe was hustled, not into the station, but into a small cement-block building with bars on the window, and, once they had deposited him in the small building, the police closed the door and locked it with a large padlock. Inside was no water or a place to sit and that was it. Fortunately, his backpack was with him and he sat on it, writing in his journal to chronicle this strange development. Occasionally, he would listen to the sounds around him or cry out loudly for help, but no one responded, and, after a couple of hours, Joe began to wonder if he would ever get out of this place. After all, they had not asked for his passport or name, so there could be no record of his incarceration. After about five hours, Joe was cold, hungry, thirsty, and tired. What to do? He didn't know--no contact with the outside world. Was this the way his life was going to come to an end? And then he thought of Dorothy. Could she help? He didn't know what she was capable of. He was torn between thinking she was unable to do anything and, at the same time, praying for a miracle.

In the middle of this reverie, a soldier unlocked the door and beckoned him to come out. He followed the officer to the station, where an American stood next to his car. "There seems to have been some sort of mistake, putting you in that storage unit, and we have registered a formal complaint. Oh, I'm Tim Beck from the Embassy, and you are?" Tim waited expectantly for an answer. "Oh, well, it doesn't matter, since they never identified you by name. All I can say is that you are lucky that woman came to us crying and telling us the sad tale of your arrest. We see very few tourists these days and only occasionally have to help like this."

"Well, thank God she got the message to you and you could get me out of that hole."

"No problem. My instructions are to take you to the youth hostel and give you the money that was passed along to us by the Tourist Committee for any inconvenience you may have suffered."

That evening Dorothy and Joe did not stay in the youth hostel, but, instead, took a room in a good hotel and enjoyed a sumptuous meal with wine and a gushy dessert. After dinner they each had a second hot bath, washed and dried their clothes, and got into what seemed like the most luxurious king-size bed they had ever slept in. Joe and Dorothy were both exhausted from their recent adventures and wanted nothing more than rest and peaceful oblivion. But still Joe thought it only appropriate to ask Dorothy if she thought this was the time to begin having sex between them.

"What do you think, Dorothy?" he asked, and she answered.

"Not tonight, Joe. We've never tried that since we've been together, so it can wait. Do you agree?" she asked in her tired, sleepy voice.

"Fine with me," he answered and within a minute both were fast asleep.

With the remaining money left from the Tourist Committee, they bought third- class train tickets to Adana on the Syrian border and set out for the remaining four-hour car ride to Aleppo-- that is, if someone would give them a lift. Traffic along the border road was clogged and horn blasts constant. Anyone would wonder why there was so much traffic from Turkey to Syria, but they quickly got a ride and soon were in Aleppo, following, or so they were told, St. Paul's trek of long ago. The driver who picked them up told them that Aleppo was for centuries the end of the

Silk Road, which brought spices and cloths from India and China and then transferred the treasures to Western Europe.

This was indeed an important stop, but they were getting anxious to get to India and were sorry they could not visit the enormous shopping center--the largest in the Middle East--or the renowned archeological ruins of the city's past glory. They did manage to sample the famous kabobs, which translates as *grill* in Syria. Almost as soon as they arrived, they were on their way to Damascus. A new problem arose when they learned that there was no road to Baghdad, but there was a bus twice a week that they could book in advance. At the bus office Joe asked how they could get there with no road. The answer came from the man behind the counter of what was the bus company. He said he was guided by the stars and, indeed, he was the bus driver and had made the trip many times, so there was no problem.

"We even provide you with a box lunch for a small additional charge."

"You mean you travel across the desert--entirely by stars?" Joe couldn't believe this latest peculiar idea of travel. But since this was the only way to go, they decided to buy two tickets the next day. Back at the hostel, Joe unfurled his traveler's checks he kept hidden in the sole of his shoe and cashed one at the bank. But with good luck the denomination, which read "five pounds," had a large dot after the "five," and in Arabic that dot was taken as a "zero," so Joe and Dorothy received change for 50 pounds instead of just five. This was a real bonanza, since their resources were already getting low. Joe did not tell Dorothy of their good fortune, partially because he thought he might be guilty of a kind of theft--if not a criminal offence, surely an ethical transgression. In fact, that thought has never entirely disappeared from his subconscious.

Since the weather was hot, it was always better to travel at night and, of course, the driver could better see the stars. So they settled into their seats with perhaps a dozen other people, and, as soon as they were out of the eastern highway, they were in the desert. The sun had set and the driver seemed to have no hesitation with his driving, despite the absence of road signs, roads, or directions of any kind. After about two hours, most of the passengers were asleep or dozing, when suddenly the bus halted. Most passengers thought this was a rest stop and began to dismount as the driver opened the door. From the bus window, Joe and Dorothy noticed a group of perhaps a dozen uniformed and armed men, all mounted on camels.

The driver asked everyone to take out their belongings and have their passports ready for inspection. Each passenger climbed down from the bus and lined up in a row next to his baggage. Another passenger offered the only explanation anyone spoke: something to do with "looking for

Jews or Israelis." After a half hour the passengers were all permitted to get back on the bus, and off they went to Baghdad.

Once in Baghdad, the pair was taken to a brand new youth hostel built by the city in an attempt to lure students to learn more about Persia and its history. The hostel caretaker was a middle-aged retired schoolteacher, who regaled Joe and Dorothy with tales of the past greatness of his country. He knew about Job and Nineveh, Babylon and the lost empires of Assyria, and more. But he could see they were sleepy and, when they began to doze, he urged them to bed to hear more another day. The hostel beds were surprisingly comfortable. There were hot and cold showers, a secure locker for valuables, and good food specially prepared as the visitor might like it.

Dorothy, Joe and friends going south by train from Baghdad to Basra

The next day, as Joe and Dorothy had their tea and breakfast, they met several other travelers: an Australian, two Germans, and an Indian named Arun—whom, for no good reason, everyone decided to call "Ari." They all agreed that the best way to proceed to India was by taking the British

Pacific ship from Basra to Bombay with stops along the way. But since there was no traffic south to Basra, everyone called for a train from Baghdad to Basra, and that's what they did. The entire group left the next day on the old train, without air conditioning or other amenities. The train car was filled with men carrying their falcons to what could have been a falcon convention somewhere along the way. It was hot, so the open windows were a welcome effort to provide relief. But once the train reached Basra, an English health official appeared and said that, because there was plague in the area, they should all get out of the city as fast as possible. This was a problem. They had just arrived, could not proceed through Iran because of the tribal danger on the border, and had nowhere else to stay. The ship was not due for several days, so some of the group caught the return train to get out of Basra, but Dorothy and Joe found a smaller luxury ship that was used only along the Arab emirates of the Persian Gulf.

The ship had beautiful staterooms, Western or Middle Eastern food, and anything else a traveler might

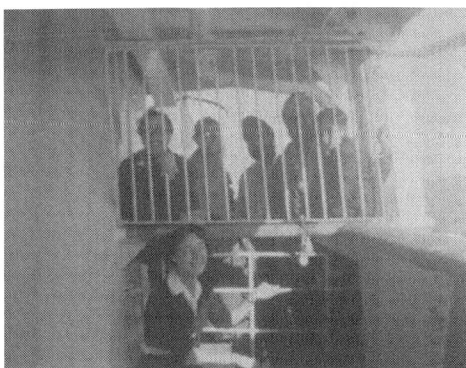

Dorothy, Ari and friends on board the ship to Bombay

51

want. It was heavily subsidized by Kuwait and usually you needed to be invited to buy a ticket. Dorothy and Joe were fortunate to have an important individual on board already. Known only as Sheik Kalifa, he was traveling alone with only one servant. Having heard in Baghdad about the pair's hitchhike-to-India plan, the sheik was intrigued and wanted to know more about their hair-brained ideas. Dorothy and Joe were invited on board, even without tickets, and gladly accepted.

The first night out was a beautiful moonlit night. The water was calm, and Dorothy and Joe ate together at a table with white linen, excellent silver, and real bone china plates. Sheik Kalifa ate alone at another table with his servant close by. At the end of the meal, the servant brought them a note from Sheik Kalifa, asking if he could have the pleasure of a conversation on deck after dinner. Both Dorothy and Joe were happy to have an opportunity to discuss whatever he had in mind, and when they met him on deck he asked about their travels and how they had journeyed thus far. He seemed vicariously to enjoy their adventures and generously told them about his studies in England and his readjustment to life in his homeland. He said that it was refreshing to hear young people who were unafraid of the world or of people very different from themselves. After an hour or so, they parted and each went to their own cabins.

"What do you make of him?" Dorothy asked.

"I enjoyed meeting him and only wish we had more time to explore his own country. As it is, I have no idea where he is from or what he does but it looks like he is a pretty big gun, wherever he is from."

The next day was pretty much a repeat of the first night, but now Joe was concerned about getting back to Basra to meet the British Pacific ship coming from Cairo and heading for Bombay. Joe expressed his gratitude for the trip but also added that he and Dorothy needed to go back to Basra.

"That is hardly a problem," Sheik Khalifa said. "The ship will just turn around and we will be back by morning. I have enjoyed talking to you and hope you will have something to remember this occasion."

After they returned to their cabin, there was a soft knock on the door and Sheik Kalfifa's servant appeared. He was invited to come into the cabin but said that was not necessary. Sheik Kalifa had instructed him to present a small gift to each of them. For Dorothy there was Rolex watch with a picture of Sheik Kalifa on the face, and for Joe the sheik had sent a small bag of perfectly matched black pearls. They thanked the servant and added that they hoped to thank Sheik Kalifa personally before they left the boat.

"That probably will not be possible," the servant explained. "The Sheik has many responsibilities and will not be available tomorrow when we dock and you board the larger ship. Oh, and he asked me to tell you that the tickets to Bombay are already paid for."

As they climbed the gangplank on the P&D ship, there were many other passengers returning to India or Pakistan. They noticed the very few Westerners in evidence. Suddenly they heard a voice calling from the deck above.

"Joe, Dorothy, up here." It was Ari, already on board ahead of them. They had to meet the bursar before finding their cabin and then joining Ari for a welcomed cup of tea in the lounge reserved for the people with upper class cabins. They were delighted to see Ari again and regaled him with accounts of their adventures with Sheik Kalifa. Ari had spent the few days just sightseeing in the area outside of Basra while awaiting the arrival of the P&O (Peninsula and Orient) ship. It was part of the British India Steam Navigation Company, which harkened back to the nineteenth century, when Britain ruled the Middle East. They spent the next week talking about India and what they were to do when they arrived in Bombay. Ari insisted that, when they arrived in India, Joe and Dorothy stay with his (apparently well-heeled) Bombay family. Ari told them a great deal about the sights and the major attractions around Bombay. One of the most extraordinary and most famous monuments was the Ajanta Caves, carved out of solid rock, with walls covered by the world's most fascinating Buddhist paintings, all showing the various stages of Buddha's life.

Ari admitted that, although he himself had never really visited the caves, he had heard so much about them that he, too, was eager to see them, sparked as he was by Joe's excitement at the thought of actually seeing them himself. Dorothy was interested, but not as excited as Joe and Ari.

After about two days on the Persian Gulf, one of the Muslim passengers suddenly and most unfortunately died. The poorer passengers were largely confined below deck and had access only to that limited area for recreation. But since a death was somewhat unusual, a small number of other Muslims and their Imam were brought up to where the deceased was to be consigned to the sea. He had been wrapped in a white shroud and prayers were recited before he was dropped into the water. Joe never understood why, but the death occasioned some agitation below and he caught snatches of shouts and demands for better rations and more time permitted outside their crowded below-deck lodgings. What Joe and Dorothy learned was largely from Ari, who spoke not only his native Hindi but Urdu as well. While none to the three had any access to the third-class areas, Ari occasionally could clearly hear—and, thus, interpret--demands shouted from below.

The clamoring eventually grew so loud that two of the ship's officers had to go below to quiet the commotion. Ari went along, and Joe volunteered as well. He had not been below and thought this the opportunity to see a part of the ship he would otherwise know nothing about. Joe keenly noted that, before going down to the area where most of the poorer passengers were contained, the two officers paused to strap on their revolvers. Nothing actually came of the incident, but it did provide a few moments of excitement in an otherwise a pretty uneventful voyage.

BOMBAY

It was a beautiful day when the ship finally docked in the Bombay harbor and everyone was excited at finally having arrived in the fabled India. The poorer passengers hurried down the gangplank, carrying their bundles of belongings, but farther down on the side, a more spacious exit was provided for the more affluent passengers, among them Joe, Dorothy, Ari, and a handful of others. Awaiting them were two Mercedes Benz sedans, each with a driver. To a well-dressed gentleman seated in one the sedans, Ari called out as he came down onto the dock.

"Dad, hello," Ari called, and his father waved, smiling. The gentleman stood up from the car and, as soon as Ari reached his father, Ari bent and touched his father's shoes as a mark of respect. Standing erect again, Ari quickly introduced Dorothy and Joe. As he did so, two of the servants ran and picked up and loaded into the waiting car, the sacks that Dorothy and Joe had carried since leaving Paris. Almost at that same instant, the captain, in his best maritime uniform, came up to Ari's father and addressed him as "Kumar Sahib," saying, "It is indeed a pleasure to see you again. If I had known your son was on board, I could have entertained him properly."

"No, no, do not be concerned. He has been at school for so many years that he has forgotten what our Indian traditions call for, but thank you for coming down to say hello. I know how busy you are with your cargo, so we will talk another time. The family is waiting to see the boy and his new friends, so we will be off." And as soon as he finished speaking, they all entered the cars and were off to the family's house in the hills.

Joe was somewhat bewildered by the reception and wondered what was coming next. When they arrived at the palatial house, they found Ari's sister and mother awaiting their arrival, surrounded by a dozen servants, most of whom remembered Ari as he was when he left (what seemed like) such a long time ago. Joe and Dorothy were introduced. Then, each having been assigned a personal servant, they were settled into a very beautiful, very spacious room. "What does this mean?" Joe wondered to himself. "Is this the India that Biro told them about?" He could see no sign of poverty or want anywhere. But he supposed every country had to have a few very rich, and he was sure they would find the real India before long.

After the reception and some refreshments, they all went to their rooms to settle in and rest. Ari and his sister Lakshmi, whom the family members all called Babs, came to see if Dorothy and Joe were comfortable. Once the young people were alone together, the two newcomers could relax and get a better picture of what to expect.

"Once you get settled in, my father wants to show you something of the real India, so take your time. You'll discover that we have plenty of time in India, so don't be in a hurry." Ari spoke sitting on the floor, his head resting against the bed, now piled with all of their belongings from their opened back packs."

Lakshmi sat next to Dorothy and said the Dhobi would take her clothes to wash and iron whatever she needed. Ari had told Babs that Joe wanted to visit the Ajanta Caves, but he was not sure what Dorothy wanted to do next. Everything was so new. While up to now they had to make do for themselves, suddenly everything was being done for them. The change in itself took some getting used to.

"Well, before anything," Joe said, "I'd like you to take me for a walk about the grounds, just to get a feel for the place."

"No problem," Ari said. "And what about you, Dorothy? Do you want to come along?"

"No, thank you. I'd just like to talk to Lakshmi...."

"Oh, forget the 'Lakshmi.' Just call me 'Babs' and I'll tell you about my years in America where I went to college in North Carolina at a place called Meredith College in Raleigh."

On their walk, Ari explained to Joe that the little huts in the back were where the servants lived and he noted how many of them had been working in the house for decades. He added that, when the servants could no longer work, some relative from their home villages would come to replace them, so the family always had a ready supply of reliable servants. When Joe asked him what the servants did, Ari explained that the Dhobi did the laundry for the family, the Malis took care of the gardens and fruit trees, the Maharajas did the cooking and prepared snacks as required. Ari's father, Adel Kumar, did a lot of entertaining, so the cooks were always busy. Another servant, early every morning, had to meet the milkman, who brought his cow to the house and milked it under the watchful eye of the servant.

"So, would say they are well paid?" Joe wanted to know.

"Paid?" Ari echoed. "They are happy to have a place to live, clothes on their backs, and occasional gratuities to send home to help support others in their village."

"That hardly seems fair," Joe protested. "What they need is a servants' union. They need to get organized to fight what looks like rampant exploitation."

Ari was not impressed with Joe's arguments, but smiled at the misconceptions of what Joe was saying. "Let's see," he said, "you've been in India for less than one day, and you are now attacking an institution that has been in existence for thousands of years. I agree it could be improved socially, and, eventually it will be, but for now neither you nor I am in a position to do more than talk about injustice and exploitation,"

When they returned to the main house, Dorothy was just getting up from her nap, and she recounted Babs's attitudes and beliefs, which had been altered by her time in the United States. She had joined the Indian Woman's Freedom Guild, which was fighting against child marriage, for more equal education for woman, and against arranged marriage, especially as it concerned her. Her parents were trying hard to bring "appropriate" potential spouses to meet her, but Babs had put most of them off with her peculiar ideas.

They talked about how long they would stay with Babs and Ari's family. Dorothy liked being in a big city, close to everything and all the amenities around them. Joe, on the other hand, believed he had come to India to find the real soul of the Continent and that this was not the place to find anything except even more of the comforts and material benefits that so choked Western life. He did not know how to explain it, but he thought that India had more to offer than Europe. Its diet was better, its clothes more comfortable, its basic values more ethical. He knew he could not really defend his ideas, but he thought in time he would become more convincing.

So Dorothy asked Joe, "I have a feeling that you are ready to get going on the search for reality. Am I right?"

Joe scratched his head and wondered if Dorothy was right. "Think so," he finally answered. And, in a sudden flash of insight, he knew what he had to do. That evening, as they all sat down to enjoy a sumptuous Indian meal, it was as if there was an electric charge in the house when Joe announced that he would be leaving the next day to visit the Ajanta Buddhist Caves and then go on from there. He thanked the family for their hospitality and warmth and hoped to see them again before his three-month visa expired and he departed from India. He left the pearls the sheik had given him with Ari's mother, to pay for a dowry for a daughter of one of the servants. A dowry often required going into debt and paying usurious interest on the necessary loans. It meant a lifetime of labor, and Joe thought it was something he could do and it might even make his introduction to India more acceptable in his own eyes.

Dorothy announced that she planned to go to the Old Portuguese colony of Goa with Ari and Babs. It had beautiful beaches, churches, and restaurants. And, after Goa, she was going to fly back to London to see her family. She had showed the Kalifa watch to Ari's dad, who had had it appraised. It came to several thousand dollars, more money than Dorothy had ever had in her life. Joe, on the other hand, still had $300 for the three months he would be in India, and he was determined to make the most of it.

The next morning, one of the cars was allotted to take Joe to Aurungabad, close to the Ajanta Caves. The whole household came out to say goodbye, including all the servants, who had somehow heard of Joe's generosity in paying for the dowry. They too wanted to express their thanks. Dorothy, Babs, and Ari all gave hugs, and Ari's dad shook Joe's hand and wished him well.

"Soon you will be on your own," he said. "I hope you will find what you are seeking, whatever it is." And with those few words, Joe's things were in the car and he was off to experience the caves. Ari's mother prepared a basket lunch for both the driver and Joe and included some samosas, water to drink, and two jelly and cucumber sandwiches.

"Be sure not to drink the water in restaurants. It will make you sick. If you are thirsty, take tea. At least the water is boiled," she warned, and, after hearing those words, they were gone.

THE AJANTA CAVES

It was a long, hot drive and took longer than expected. The driver dropped Joe near the entrance of a cave guarded by two large, carved, stone elephants. Although it was late, the driver insisted on returning that night to Bombay, and, after helping to get Joe's bag out of the car, he was gone, and Joe stood there, looking up at what seemed a gigantic carved stone cathedral. Of course, it was no such thing, but Joe was transported at being in the presence of what he felt was going to be the most wonderful experience of his life. There were no tourists at this hour and the guards were nowhere to be seen, if there had been any to begin with. So Joe picked up his bag and walked into one of the rooms behind the sculpted and decorated facades. Many had been cut from or into the volcanic basalt rock over 1500 years earlier--by men using only hand tools. Although the paintings were deservedly better known, the rock carvings were a world-class demonstration of the devotion of the craftsmen of that time.

The interior of the room Joe entered was in utter blackness, so he settled just under the entrance where the starlight could help him identify a little of his surroundings. It was a strange and different experience for Joe--being totally alone with no knowledge of his location or even why he was there. But instead of fear or foreboding, he felt protected, comforted, as if he belonged there and would be richer for the experience. And with those few thoughts, he fell fast asleep.

Ajanta Caves

He awoke before dawn and wondered where he was. The air was cool, and to the east he could just begin to make out the approaching dawn. He looked into the basket Ari's mom had given him and found the samosas and a cucumber-and-peanut butter sandwich that tasted delicious. She had put in some more chilies because she knew how much he would appreciate the extra spice. After that early breakfast, he took a walk around the caves to see what he could, and then sat down where he had slept and thought for a moment. It was not exactly a meditation, but he thought about what he was trying to accomplish at this moment in his life.

After all, he was not working and not thinking about work. He had little money but was quite unconcerned about having more. In fact, he was comforted by the thought that he seemed to have overcome dependence on accumulating goods and cash, and, after all, wasn't that one of the ideals that Buddha preached? Of course it was, and now he felt it was falling on him to realize the teachings of the great sage. But at the same time, he knew enough to realize that he was not free of the dependency on the world. He had to eat, he looked for a life with a woman, and he wanted to improve the world. Wasn't that in itself a burden that would hold back his freedom from dependency on others?

Ajanta Caves

He thought of the Christian doctrine that he had gleaned from sporadic contact with the Catholic Church. There they thought of going to "heaven" and were told how to get there through prayer, charity, and love, but the Buddha had no such ideas. The Buddha taught his followers to give up craving, whether it was for ethics or anything else, and to forget about heaven. All life was wanting something and suffering, since there is never enough and always more to add to what we have. For Buddha, the good was to avoid harming others, forgive your enemies, or, better still, to have no enemies.

All this he had read but it had sounded like boilermaker prose, but now Joe wondered how it applied to his life. He had just left Dorothy, and when they parted she whispered that she loved him. And didn't Lea say the same thing? So was he in love, sitting here at the Ajanta Caves and wondering where his life was headed? He felt the power of the Buddha and his great knowledge of how our attachment to our petty lives would keep us always from attaining fulfillment. As these thoughts floated in and out of his consciousness, he noticed that the sun was already penetrating some of the darker recesses of the caves. That meant that sooner or later the tourist buses would arrive, filled with their largely Westerners searching for Enlightenment, or just fun.

The good thing about the visitors was that they would pay for electric lights, which enabled anyone to see the glorious frescos and statues that were otherwise lost to sight. And just at that moment, as the government guard came to life and appeared as if he had been on duty all night, the tourist bus stopped at the cave entrance with a pile of busy people all decorated with sun hats, cameras, flowered clothing and little lunch boxes the hotel had given them, together with bottles of water. The electric lights miraculously turned on as the crowd gathered together around the tour guide to hear the story of the caves and the life of Gautama.

Ajanta Caves

Joe, happy to get a place in the back of the crowd, walked into the various caves with everyone else. The light added a dimension and it was wonderful to see the hidden treasures where no daylight could penetrate. Joe wondered how the carvers could have done such exquisite work in the near darkness. After an hour or so, the guide told the people they could rest or go back inside the caves on their own. There were also a few tables with umbrellas where some opened their box lunches and checked to see what had been put in for them. One woman opened a conversation with Joe, asking, "I didn't see you on the bus. What hotel are you staying at?"

"I'm not at a hotel at all. I just slept on the temple floor last night so I could see inside the caves when your tour group came and paid for the lights to come on."

"Are you some kind of mystic?" she asked.

"No, not at all, but I am here to learn more about Indian culture and life. And what about you? Are you here to learn about Buddhism?" Before she could answer, another woman at the same table complained loudly.

"My God, I can't believe they put the same lousy sandwich in my lunch box after I told them I do not eat eggplant. I am going to throw this away. After all I paid for this trip, we should at least get a decent lunch."

After she spoke, Joe asked the young woman if she wouldn't mind passing the woman's box over to him, which she readily did. "That will make a dinner for me this evening."

"So you really are on a tight budget," she said.

"Well, you could say that, but I do not like to waste good food when there are so many hungry people around the world. Do you understand what I am saying?"

"I guess so," she answered uncertainly.

"So what is your story?" Joe asked, more out of curiosity than concern. She wore a pair of denim shorts, a plaid jacket, and a broad brimmed straw hat, from the back of which hung her ponytail.

"Well, I'm a student and can get some extra credit for taking this trip. My friend Bertha warned me that all the others on the party would be ill or infirm and there would be no people my own age on the trip, but I said it didn't matter. In fact, you are the first person near my age I have talked to since we left Toledo, Ohio. Oh, that's where I'm from in case you want to know. McIlhaney High School. We almost won the state women's volleyball championship, except that the referee liked our rival a lot more and unfairly gave them the final game and we came in second."

"But, anyway, are you getting something out of the trip?"

"I guess," she answered. "Oh, my name is Shirley. And what I like is the choices they give you. For example, we could visit the Ajanta Caves here or go to Sevagram, which was sort of a school started by Mahatma Gandhi, but no one wanted to go there when we learned that they didn't have hot water or showers, you couldn't lock your door, and had to sleep on what they called a Charpai, which the tour guide said was a kind of rope bed without a mattress. Well, I wouldn't mind, but I was the only one interested and they couldn't make the detour for just me."

"So you wanted to kind of rough it at Sevagram, is that right?" Joe asked.

"Well, though I'm not complaining, I do feel we are treated like a bunch of sheep. We get up at a certain time, catch the bus, and spend so many hours doing whatever, whether we're interested or not, and then eat and off to bed. The trip costs an arm and a leg, but it is not at all customized or made to fit our schedule. I thought about you setting your own pace and costing you nothing and I wished I could join you."

"But you know there is a down side. Sometimes it is too cold or too hot, you may not have anything to eat, and wherever you are going you may not get there for days." Joe looked at Shirley to see if what he was saying was getting through to her.

"I understand," she answered. "But if I could spend just one night here, without the bus tour people, I could really have something to impress Bertha when I got back to Toledo. To me it would be a real coup, sleeping in the presence of all these Buddha and waking with the sun streaming. Oh, it just takes my breath away just thinking about it."

"It's good to hear you speak about this place so meaningfully. I feel the same way," Joe said. "But I don't think your motives are to be trusted. You belong on that bus with those people. You are an American and, as such, pretty set in the virtues of that culture which includes consumerism, acquiring more things, being unhappy with your choices, and feeling you are better than most others on the face of this planet. You should go back and not worry about the Ajanta Caves or Buddhism."

"So you think you can make choices for me, do you? Well I'll decide which way to go, thank you." And with that final word, she picked up and went off to see where the others were headed.

After the bus honked for everyone to be on board, it took off for the trip back to Aurungabad. The driver assumed that Shirley had joined another party getting back to their hotel. Joe went off to look at the inside of a cave he had not explored earlier. The interior lights were on a timer and would not go off for another fifteen minutes, so all the glory of that cave was his alone to enjoy. Except that he was not alone, for a person came out of the back from behind the great statue of a sleeping Buddha and, lo and behold, it was Shirley.

Joe greeted her, "So, you decided to take the plunge to have more to tell Bertha."

"Well, sort of," Shirley said, smiling. "I guess you and I will be spending the night here."

"I can't think of a better place to sleep than here."

As the sun went down, Shirley and Joe sat together, eating the remains of the box lunches Shirley had thought to collect from a few other passengers in her group. They enjoyed eating and talking about their relatives and the neighborhoods where they grew up. Soon it was time to sleep and, since Joe had a light cover, he used it for both Shirley and himself. It was not large and Shirley had to be pressed next to Joe so it would fit over both of them. Joe had not been with a woman since he left Paris and was surprised at how the prospect of enjoying Shirley kept coming to his mind, so he turned to see if the same feeling was with Shirley. As soon as he touched her body, she readily removed her under things and helped Joe get into her. After that, they nestled closely together until the sun was just beginning to rise. The colors of night were passing and a new day was dawning.

"Did you enjoy last night?" she asked.

"I certainly did," Joe answered. "More than I expected. And you?"

"Oh, God, yes," she answered, "and I want more before the bus arrives, but, first, where is the bathroom?"

"There isn't any, but you see the scrub of weeds down the hill. That's where you can go, and then it's my turn." After they were both comfortable and had had more sex, practically under the eyes of the Buddha, Joe said his prayers or meditated, and Shirley just sat, enjoying the morning and reminiscing about the time she had spent with Joe.

SEVAGRAM

When the hotel bus arrived with new tour passengers, Shirley spoke to the driver, who agreed to take Joe back to the Aurungabad station where he could get a train to Wardha, which was as close as a train went to Sevagram. The wait was long, but eventually they reached the station. Joe just hopped off the bus while Shirley looked away, not wanting to experience the departure. There were no buses to Sevagram for several days, and it was not clear exactly when one was expected. Joe had to decide if he wanted to wait, sleep at the train station with several other passengers on the cement floor, or set out walking n the direction the station master indicated to get to Sevagram.

"There is another alternative," the station master offered, "but I'm not sure it will be to your liking."

"What is it?" Joe asked eagerly.

"You see that bullock cart, picking up some materials; he will deliver to Sevagram in the morning. It will travel all night to get there and you can sleep as well as the driver. The bullock knows the way and by morning you will arrive. The bullock moves about as fast as you can walk, slowly, but it will get there by morning." Joe thought about it and decided it was better than sleeping on the floor or waiting for a bus that had no scheduled arrival or departure. The station master spoke to the driver and told Joe to give him a rupee now and another when he arrived. There was straw in the wagon. Joe made himself comfortable and suddenly realized how tired he was. He had been sleeping on the stone steps of Ajanta and had forgotten what a nice straw bed could do to rest your weary bones.

As dawn showed the outline of the thatched buildings that was Sevagram, Joe saw people already moving about in the little village. There did not seem to be any central office where Joe could report his presence, but several people arrived to help unload the few supplies that were being delivered. As Joe arose from the straw and was recognized as a foreigner, it caused a stir in the village. There had not been many visitors since Gandhi's death, and, even before then, most tourists, warned about the intense heat as well and the prevalence of snakes and scorpions had shied away from visiting the village.

The first words Joe heard were, *"Aap Kaha se atee hain?"* [Where are you from?] Spoken by someone Joe assumed was a member of the ashram. And when there was no answer, they just smiled and called for another member who knew English. "He is from America," the woman spoke in Hindi to the people, who had now gathered around the bullock cart and the new arrival. "He is interested in learning about education and living within your means." Joe learned only much later what was going on, but was happy to have someone who could explain his purpose to others.

Sevagram

The educated young woman quickly took charge of Joe and led him to a hut, where he met someone who seemed to have some authority, a man named Naren Tambe. Nina was the woman already taking Joe to meet people. He was assigned a string bed, a charpai, and told to get ready for morning prayers, a daily ritual in Sevagram. There may have been a hundred people who showed up at the clearing in the center of the village. Tambe led the singing and prayers, but all Joe could distinguish was the names of a few deities.

"Jai Sita Ram," they sang, and soon Joe was taking part as fully as he could and trying to bow with the others, stand up, clasp his hands in unison, and join again in "Jai Sita Ram."   After the morning prayers, groups formed for breakfast, and Nina and half a dozen others shared tea, japatis, and, this day, each had a banana. Joe felt pretty good and was learning some Hindi. Each new acquaintance began by clasping their hands and saying, "Namaste," which is a sort of "Hello and welcome."

Nina explained to Joe that normally they ate off leaves so there were few dishes to wash. But when they had to wash dishes, they used the ashes from the previous day's fires as a kind of cleanser. After breakfast everyone had a job to do, since the idea of the ashram was to make everyone self-sufficient by growing vegetables, spinning cotton, growing cotton, weaving cloth, or some other activity. Nina asked some others to take Joe with them into the fields to pick cotton and then prepare for the next step in the eventual goal of weaving clothing. Most people wore cotton--the women, saris; the men, dhotis. The saying was that if you learned to spin well, you had a generous garment, but, if not, yours remained short and meager.

By noon the sun was already hot and they came from the fields to rest, bathe, and have lunch. The lunch was rice and curry and *alu* (potato), which Joe enjoyed eating now with his new-found buddies from the field. They wanted to learn English and Joe wanted to learn Hindi, so they traded words, although this was not exactly Gandhi's idea of what he called Nai Talim, or New

Education. Remember, Gandhi was interested in reform and freedom, in getting rid of the English tyranny, so he emphasized a return to native culture and less reliance on England. The amazing thing is that he succeeded in getting rid of the British, but not the English language.

Everything that Joe encountered was a delight. He learned to spin cotton, first with the takli and then on the spinning wheel [charka]. He learned some rudimentary Hindi and found that Gandhi's ideas contained the values he would cling to the rest of his life. One of the fondest memories he kept is of group spinning after dark. Dinner was over and a few candles lit in the main hall where whoever wished could join to spin. This was a place where people came from all over India, and, while they spun, some would sing their spinning songs. A few went like this:

> *Ghoon ghoon, O my spinning wheel,*
> *Should I spin the red pooni or not?*
> *Spin, my girl, spin.*
> *Distant is my father-in-law's place.*
> *Should I go and live there or not?*
> *Go there, my girl, go there.*
> *Very long is my woeful story.*
> *Should I relate it or not?*
> *Relate, girl, relate.*
> *My husband is really a minor.*
> *Should I stay with him or not?*
> *Stay my girl, stay.*

Someone asked how the singer remembered these and many more songs she sang. The woman replied, "It is the spinning wheel itself that keeps my memory fresh. It gets each word of the song again and again etched on my mind. It makes every word alive." After one stopped singing, another would pick up another song.

My love is a lump of candy.
He talks so sweetly.
My love is a cypress plant.
I got him from God.
I feel lonely while spinning.
Ask my husband to come home.
Your drunkard son, O widow,
Broke the axle of my spinning wheel.
Hearing the sound of the spinning wheel,
The ascetic came down from the mountain.

Spinning remained an important activity in many homes and villages and was considered a part of a bride's accomplishments before marriage. There are many stories about spinning, which are passed around as people gather together to make the yarn. The Sufi poet Hussain (1539-1593) could compare his soul to a girl who remained unmarried because she failed to prepare her trousseau with yarn as the ancient tradition dictated. She had to do this herself and when she failed, she remained unwed. The spinning, the songs, and the personal reminiscences that filled the quiet night, all continued and were a welcome kind of relaxation and learning.

After two weeks, a visitor arrived, who was already nationally known as an incarnation of Gandhi. He was Vinoba Bhave, who went from village to village in Bihar, Uttar Pradesh, and elsewhere, asking the large land owners in the area to improve the condition of the poor by breaking up their holding and giving the needy pieces of land for them to own and cultivate. Naturally, this was not an easy sell, but Vinoba had already achieved considerable success and was now looking for more recruits to help him with expanding his mission. He was heading for Puri in Orissa, and, when Joe heard of his daring mission, he sought to join him and was accepted.

Bhave's aide said to Joe, "Although you know no Hindi or any other Indian language, your presence may help by showing that out message of sharing the bounty of God with our least fortunate brothers and sisters has spread, even to the great country of America." So Joe packed his bag and joined others from Sevagram heading to the Puri Sammelan meeting in the town of Puri for activists. At one of the sessions where he could not understand a word of what was going on, Joe went to the beach to swim instead. But the waves were treacherous and Joe was afraid to go into the water. Recognizing his hesitation, a young man came along with a rope and offered to tie it around

Joe's waist so, if he was in trouble, Hari (that was the prospective life guard's name) could pull him out of the water. It seemed like a wonderful idea and, sure enough, as the waves knocked Joe down and he was unable to get up from the strong waves, Hari would pull him to shore. Joe felt as if his life had been saved and thanked him profusely and gave Hari a generous tip for his service.

SIVANANDA

Back at the Bhave enclave, Joe met a Brahmin who asked him if he had come to India to learn about himself.

"I think you're right," Joe answered. "Originally I came to learn about India and to feel comfortable with less. But I have forgotten about that and have found so much more to feel and enjoy."

"That is good," the man said, "but you have only scratched the surface. Do you think you are ready for some serious introspection? If so, I will tell you where to go next."

"You don't know me from anyone else," Joe, good naturedly challenged the man, whose name was Parikh. "How do you know you can advise me and be helpful?"

"I cannot answer your questions in a way that will seem logical to you, but I will tell you anyway, and you will do whatever you wish. You must go to visit Swami Sivananda at Sivananda

Nagar, near Rishi Kesh." And with those words he disappeared and Joe never saw him again.

But his words left a nagging doubt in Joe's mind. "Should I go back to Sevagram or visit the Taj Mahal like every other tourist--or try to find Sivananda?"

He stood a long time, just looking out at the breaking waves along the Orissa shore. Then he returned to the Bhave compound to take his leave and find a train to Rishi Kesh, the nearest train stop to Sivananda Nagar.

Early the next morning he bought a ticket for a third-class seat on the train, headed first to Delhi and then to Dehra Dun in the beginning of the Himalayas. The train was expectedly crowded, with rows of benches on each side of the cabin; all taken up with so many more people crowded and sitting on the floor that there was little room, even to stand. What was unusual was that one of the two benches in the cabin--which might easily have accommodated six or eight passengers--, was completely taken up with a single passenger, lying fully stretched out on the bench while women and children stood or sat on the floor. None of the other passengers moved to ask the single passenger to move, or, rather, they all accepted the inequity as normal. "Don't some always have more than others?" they may have thought. But not Joe! Without a word of warning, he just pushed the man's feet and legs off one end of the bench and sat himself down. The man was surprised but said nothing and only gathered his things and moved to the other end of the bench while a few other passengers easily moved to sit in the space between Joe and the lounger.

The trip to Delhi would take all night and all day, and at each stop there were sellers with tea, sweets, samosas, and more. These, they could hand to the passengers through the barred and open windows. Everyone had to be fast before the nimble monkeys could grab something for themselves. But eventually, tired and hot, Delhi arrived and Joe stopped for a hotel stay so he could rest, eat, wash his clothes, and take account of where he was going. After a couple of days, he felt refreshed and still had enough money to last him to Calcutta and a ticket to Japan. At least that was the plan. For the moment, he was content to head for the mountains and Sivananda Nagar. Once he arrived, there was still a bus to reach the area he needed. And there were still several kilometers to the actual ashram. Fortunately, a horse-drawn wagon waited at the bus stop for the occasional visitor. This day, Joe was the only visitor.

The road ended at the ashram, where a young man wearing a long skirt asked Joe in Hindi what he was looking for.

"*Aap Kya Chaiyee bhai*," [What are you seeking, brother?]. All Joe could answer was, "Swami Sivananda," which only revealed that the newcomer was not well acquainted with Hindi. So the young man continued in English.

"Sivananda is cleansing himself in the river, as he does daily, but I will tell him later you wish to speak to him. In the meantime, I will assign you a room and some refreshments, if you like."

"That will be very nice," Joe responded and was led up to a well-built stone building like many others spread up and down the steep valley leading to the river where Sivananda was bathing. Once in the large room where he was to stay, Joe noticed the heavy wooden doors and heavy brass hardware to keep them securely shut. There were no windows in the room, but high up near the ceiling were barred openings that you could look out of if you were tall enough. Soon, another man came to the door with a tray of fruit, some japatis, and papaya juice. He carried a stick, which he used to fend off the monkeys, who sought to take what they could from his cloth-covered tray.

Outside, the monkeys hung onto the bars in the windows and screeched as if to say, "You have something and we have none. Share you, stingy devil." But as Joe ate, they soon disappeared and he was left in peace. The man who had originally met Joe returned to ask his nationality and where he had come from. When Joe mentioned Vinoba Bhave and the Puri Sammelen, his face brightened and he said, "You are the first American who has come this way since the end of the war, and, certainly, the first who has heard of Vinoba's work."

"Well, actually, I did very little except learn about how poor some people are and how wealthy others remain. But what is most interesting is the fact that under Bhave's urging many land owners actually found they do not need the wealth they have carried like a heavy burden around their necks. And they actually have found themselves happier giving rather than getting more. My fellow Americans would never believe that is possible, but, if I ever have the chance, I will tell them."

"Good for you," the man answered. His name turned out to be Prasad, and he said he had a few things Joe should know about Sivananda Nagar.

"First," he said, "you will notice we have no electricity. Occasionally we light candles, but generally we go to bed at night and get up at dawn. Food is modest but plentiful. We drink juices and eat yogurt, rice, Nan, fruit, and vegetables. Rarely will you find salt, sugar, onions or peppers. You can remain here as long as you wish. There will be opportunities to learn about the wisdom of Sivananda. For example, tonight many people will come to ask Sivananda for a mantra to live by, and you can ask for one yourself. If you need anything, just ask for me. Just say in Hindi, 'Prasad, kaha hai?'"

"Thank you," was all Joe could say. As soon as Joe arrived at the ashram, he bathed in the Ganges. The water was wonderfully cool and refreshing. Steps led down to the water. From there the inhabitants washed themselves or their clothes. Joe was a bit hesitant at first, but, seeing everyone else doing it, he undressed except for his shorts and plunged in. The water was too rapid for swimming, but for dunking it was fine. Joe managed to wash all his clothes and lay them on the rocks to dry. In an hour he put them on again and walked off--by far the easiest wash he had had in India thus far.

That evening after dark, a line of ashram residents and visitors arrived to seek personal mantras from Sivananda. They lined up outside the simple house he himself occupied and waited patiently for his appearance. This ritual took place whenever Sivananda was in residence at the ashram. The mantra was a sort of adage or proverb an individual could use as a guide for his personal realization. The night Joe attended the mantra session was cool and the area lit only by abundant starlight. Most participants wore shawls, but, since Joe had none, one was given to him by Prasad. The line moved quickly as Sivananda barely looked at the individual before pronouncing the guiding principal that would last for the rest of that person's life. When Joe first saw Sivananda, he was disappointed. He had expected (from the photo in the paper Biro had shown him) a very tall, thin, ascetic looking man. Indeed, he was tall but large, what you might even call corpulent, not unlike some Italian monks whose monasteries have become more famous for their fine wines and excellent cuisine than for their spirituality. The swami walked with difficulty, which necessitated a cane, giving the impression of some muscular ailment.

From earlier photos, which still grace the majority of his publications, he had suffered a marked deterioration. Whether this was due to the life-giving potency of his medication or not, Joe could not say. That evening Sivananda shuffled slowly onto the open patio of the ashram, where the disciples had gathered. He took a few steps and one of the orange robed sadhus prostrated himself full length in front of him. Instead of receiving this homage with the august sanctity, which it could easily have provoked, he seemed not to notice, except to call out the sadhu's name and shine his flashlight into the fellow's face.

After everyone had bowed and scraped sufficiently, the Swami eased himself heavily into the padded armchair prepared for him, and the evening meeting continued. From across the river came the faint sounds of bells and singing. Sivananda ashram is not alone in Rishikesh, nor was it unique. Only the Swami was different.

The meeting progressed with the Swami calling the names of various sadhus and disciples, who mounted the low platform in front of the assembly and there either sang, prayed, chanted, or discussed philosophy in Hindi or English. What was amazing about the Swami was that nothing was made sacred. The novitiate could assume all of his self-righteous ardor or begin a prayer, but the Swami would begin joking about him, saying, "My, we're looking austere tonight." And the assembly would laugh, including the sadhu, despite himself. Everyone seemed quite used to this sort of heckling and joined in. There was nothing gloomy or sacred about the ashram; hence, it had a rather healthy and happy atmosphere. After the puja and flame worship to Shiva, the Hindu temple sanctity came upon the ashram and Sivananda accepted the mantle of divinity.

When Joe approached Sivananda that night, Swami had already been informed about the newcomer and asked bluntly, "Why do you want a mantra?"

"I'm not sure," Joe answered honestly. "I just thought, since I have come all this way, it seemed like a good idea."

"Many of the people who come to me will realize their fulfillment with the help of the mantra they receive today. But for you it is different. You are an American and eventually will find yourself back, perhaps in New York, where you may stand on a street corner mumbling something like 'Om Mani Padme Hum' and everyone around you will think you are peculiar, and, indeed, you yourself will find your own action fantastic. No, at this time, this is not for you."

Joe was taken aback by the frankness of Sivananda's remarks and only stood perplexed before moving away so that others could come forward. Joe felt depressed by the encounter and could only wonder if he had made a mistake coming here in the first place. As Joe moved up the path toward the room assigned him, Prasad called him to come into his own room so he could talk to Joe. In the room there were no chairs, so they both sat on the floor and Prasad brought out some Nimkin [saltys] to nibble on while they chatted.

"Don't be unhappy with what happened tonight," Prasad said. "Many people are turned away, only to return the next year and have better luck. I am sure you know little about Sivananda, although he is quite famous in some ways. He travels widely, publishes books, many of which are sold in our bookshop, and he is invited often to visit other countries. A few years ago he met the Queen of England and instructed her on Hindu meditation, and she promised to return the visit by coming here one day. Many people have visited this sacred place and often made considerable donations for our work. As a result, this ashram is very well endowed and financially rivals some of your better colleges and universities. This is not to say we look for donations. Anyone like you can remain here at no cost, so long as he is a seeker of the divine. So please do not be discouraged. In fact, Sivananda mentioned to me that he saw promise in you and intended to help you."

"Well, what you have told me is great news, and I look forward to speaking to Sivananda whenever he calls me." And with that Joe returned to his room and slept soundly. The next few days Joe spent wandering over the glorious hills and speaking to a few others who were also in residence. Then one day, after the evening meal, there was a knock on the door and Joe was shocked to see Sivananda himself.

"May I come in?" he asked

Joe was so surprised to see Sivananda himself that he could only stutter, "P-P- Please do."

"So I understand you are interested in meditation, Hinduism, finding your way into the nature of the divine and such things like that. Is that correct?" he asked.

"Well, yes," Joe answered. "I don't really know what to say or how to frame a question, which is so difficult and impossible to describe."

"Let me give you some guidance in this," Sivananda said. "I will draw on some of my own experience, which may help you a little or a lot. Each individual is different, but we all seek the same thing. Start with looking at the mountains and expect nothing. Clear your mind of as much as you can; empty it like a pitcher of water so not a drop remains. Breathe deeply and think of nothing. This is not easy, since normally people are full of ideas, expectations, hopes, and dreams. So to clear the mind is a vast accomplishment in itself. But be patient. You have spent a lifetime learning about yourself and others, and now you are seeking nothing. Wait! It is the most important thing you can do. Wait.

"After some time--it may be an hour, it may be a day, it may be a lifetime or longer—but, if you persist, eventually you will forget yourself. When that happens you will cease to exist. And you will merge with a great cosmos that is nameless. You will enter a divine universe about which you now know nothing and never will. It is an experience beyond yourself in which there is neither you nor anyone else. There is no life and there is no death."

Joe listened, spellbound, but still felt he had to ask, "So are you saying that meditation reveals a world without any gain for the life we live here and now? It sounds like a lot of effort for very little benefit."

"You are quite right," Sivananda answered. "Seeking the divine is not for everyone. It is difficult and, when you have reached Nirvana, there is no pot of gold nor is there what you might call success. It is rare for some great sage not only to find fulfillment but also to find answers to the great problems facing humanity. But it does permit you to experience the shadow of what you went through and, with that, you are aware that others feel in your presence the journey you have taken. So we have said enough for now. See what you can do while here and then see The Mother in Pondicherry."

And with those words, Sivananda was gone, not only from Joe's room but from the ashram itself, headed to a lecture at Tokyo University. Joe's only remaining thought was, "Who is The Mother and where can I find her?"

PONDICHERRY, THE MOTHER, AND AUROBINDO

It was not difficult to learn about the Aurobindo Ashram in Pondicherry. Every school child in India knew the history of how this revolutionary, Aurobindo, had first served the British and then the Maharajah of Baroda, was arrested for being a revolutionary, and finally settled in the city of Madras, where he and The Mother jointly ran the most unusual ashram in India. But before going south, Joe tried to follow Sivananda's advice and meditate into oblivion.

He arose at dawn to watch the sun come over the mountains, sat with crossed legs, and relaxed his hands in his lap and tried to think of nothing. He tried earnestly to empty himself of everything, but, as hard as he tried, he never seemed able to eliminate his own self-consciousness, of the sense of what he was trying to do. He could never get away from himself. So, after a short while, he decided that this was going to take longer than he thought and made plans to head for Pondicherry.

It was a long trip by train, from one end of the country to the other, with occasional stops to rest and recuperate. He had only a few weeks before he planned to be in Calcutta for the boat to Japan and, eventually, to the United States. But for the moment his goal was Pondicherry and "The Mother." What awaited him was still unclear, but, since Sivananda had pointed the way, Joe would never refuse that call. Three days after leaving Dehra Dun, Joe was walking into the Aurobindo Ashram and was greeted by one of the officials, who did not ask him what he wanted or why he had come to Pondicherry. Instead, he was given a room and a schedule of when meals were served in the cafeteria. In the ashram was a large publishing enterprise--perhaps the largest in India at the time, which published books in all the Indian languages. It also contained a school for several hundred neighborhood children, who were always conspicuous in their bright blue, carefully starched uniforms.

But the first words he heard from the man who showed him to his Western-style room was something like, "We have no rules for reaching salvation. That is all up to you. We have yoga, meditation, and prayer, but that is for you to choose. You must find your own way. Take as long as you like and we will help to do what we can. Anything you need, just ask."

Joe was delighted to see the desk and chair, the bed he had not seen in some time, and he ventured to ask if perhaps there was a typewriter available that he could borrow.

"I will see what I can do," came the answer. "And as far as Mother is concerned, she is not strong, but you will see her on the tennis court before dinner. She rides her car to the court and sometimes hits a few balls. When she was in better health, she enjoyed the game a great deal. After dinner she usually reads a story to the children who reside in the ashram with their parents, so that will be another good occasion for you to see her. Before tea time at 5 PM, you can just wander around and become familiar with our facilities."

From his window, Joe could look north along the shore and see the white breakers along the Parc de Charbon shore. Each morning, a pitcher of water was brought to the room with an offer to wash his clothes. Joe was happy for the offer, but, since he was still able to do it himself, he preferred it that way. The old quai built by the French, was still there, but a storm had broken it in half and it could not be used at all. Apart from some fishing, there was nothing in the town but the ashram. One day Joe met a medical couple at dinner. Both of them had come to see The Mother, who they both believed was a genuine spiritual force. At dinner, the doctor recounted the miraculous story of how The Mother had forbidden all ashramites to go on the jetty just previous to its collapse.

"You see," he said, "there is some power marking here. These are not ordinary things."

There were hundreds like him, accepting the Mother as a supernatural being and worshiping her as divine. For example, at the library, Joe met Stephany Pomeroy, an American. In conversation he told her why he wanted to see the Ashram. He had heard that by following Aurobindo's philosophy, people learned to live in this world and simultaneously in the next.

"Not only that," Stephany added emphatically, "but to bring that next world down here now and make it a part of our lives." Another American at the library wore an orange robe and a rosary with the Mother's picture around his neck at all times. He said he had surrendered himself completely to the Mother. She had built him a small house a few miles from the ashram from where he cycled in each day. After two years there, he was ready to spend the rest of his life. "Why?" Joe asked him.

He answered, "To find the living presence of God within me, to make it transform my life, my body, and every drop of my blood to the divine way, and it can only be done by The Mother's grace."

His was not an isolated case. Everyone there, from the children to the most aged, all fell prostrate before her, seeking her help and guidance. When Joe arrived on Sunday, he was hustled off to the Ashram Hotel, an expensive, Western-style place which was to serve temporarily as his lodging. The next day was Monday and the Mother's birthday. A special blessing was arranged for everyone, and when they arrived at 9 o'clock many people were already waiting, praying, and meditating around the Samadhi of Aurobindo, a large concrete tomb, simple but decorated with every kind of fresh flower arranged in patterns and covering the entire surface, perhaps 10 feet by 5 feet. A group of women arranged the flowers beautifully in the most intricate designs. Removing their shoes, the group mounted through the rooms used by Aurobindo for the last forty years of his life. From there they entered a large hall where the Mother sat.

As he entered the large hall, Joe found perhaps fifteen people ahead of him in the line. Her face was heavily powered and on her lips she wore a line of red lipstick. As he walked into the room from the far end, Joe could see her darting eyes and a grimace fleeted across her face. Her sweet smile returned and she wore it as he came up to her. She handed him a poem. He was afraid of her absolute autocratic control of the ashram and everything in it. This fear is what prevented him from seeing her other qualities, such as mercy and generosity. He related his feelings to a Moroccan Buddhist he met there named Ben Simon. Ben

Simone's feelings were quite different. He said of her, *"Plein de force et magnetism."* So Joe went back to practicing yoga.

The routine of the ashram was quite simple. Daily, everyone got up in time for the Mother's *darshan*, or balcony blessing, at 6:15. The community gathered in the small street below where she lived. They waited for her to appear. She wore white silk, and, when she appeared everyone, rose silently. She then surveyed every face in the assembled crowd, and silently disappeared. After that there was breakfast, always the same: cocoa, a small loaf of whole wheat bread, and two bananas. The dining hall was a series of light, airy rooms with tile floors and tiny individual tables. It was a wonderful feeling to walk barefoot over the cool tiles and squat and eat. Joe enjoyed these meals enormously. For those wishing to work, there was concrete block building, a huge metal factory, and several buildings devoted to printing books. Joe asked one ashram worker how many hours a day he worked. "Eight, ten, twenty," he answered. "What does it matter? It is not work; it is my life that I am living."

Another worker volunteered, "It doesn't matter how one lives--whether poorly or richly. The thing is not to be attached to one's wealth. If a man has five rupees and is attached to it, he will not progress spiritually, but if a man has five lakhs and is not attached to it, it will not hinder him."

100

The great liberty of the ashram allowed people to come and go as they wished. They were permitted to have servants, or to take tea, or have spices if they wished. Many lived with their families and did not surrender completely. Even people who did not consider the Mother divine were allowed to remain there.

Joe lay on the bed and marveled at how little he missed the clean, ironed sheets he remembered from so long ago. Almost asleep, he heard a knock on the door and the typewriter he requested was deposited on his desk with paper and ribbon, ready to go.

"Thank you very much," Joe said. "But how do I pay for all this--the room, the food, the typewriter?"

"You are our guest," the man said. "And for guests there are no charges for anything we have to offer. You will be surprised to know that an American came here a month ago and said he always wanted an island where he could live entirely alone, and Mother gave him his heart's desire."

"This is hard to believe," Joe said. "How can you do that with so many different people?"

"It is the Mother. She does it all. We really have no administration. It is all done in some miraculous way. If you stay awhile, you will be amazed at how things just happen to get done. There is a Raja who came last year with his elephant and servant, and he is quite content in his new living quarters. He has turned over all of his worldly goods to the ashram so that others can benefit as well."

As soon as the man left, Joe began writing up his experiences and the philosophy of the Aurobindo Ashram. It took a while for him to fathom how it was possible to run a school and a publishing enterprise to accommodate tourists and scholars and many more seekers of truth--and do it so effortlessly. That is the unbelievable part. That evening Joe attended a reading in a large living room in the Mother's residence. A dozen children were present, and The Mother sat in a large comfortable chair and read a story in French. It was the Mother's native language and these students were all learning to read and write French. The story, called *Le Viellard Qui Fait Fleurie Les Arbes Morts,* concerned the magic that awaits true believers to overcome aging and even death.

The Mother's voice was not strong but very pleasant, and the children had to pay close attention to every word in order to understand the story. French was, after all, their third language. They all knew their mother tongue and had learned English since infancy, and now a third language was being used and they all seemed to want to learn as much as they could. It was so different from some schools Joe knew in America. He delighted in witnessing this wonderful learning.

After the reading, everyone went home, and Joe wondered if he might have a chance to express his appreciation to The Mother for all she had done for him. As he went forward toward her, she rose and faltered, needing help from two attendants, who were poised nearby in case of just such an emergency. So Joe returned to his room to make plans for his next trip. It was already late and his visa had only a brief time before it expired, but, more important for him, was the fact that his money was nearly gone. He had started with $300 for three months but had to save $100--for his passage from Calcutta to Japan and then, where? He didn't know it yet, but, before Calcutta, he had to stop at the art school at Shantiniketan and also visit a Santal village and then see the religious life of Calcutta through the gods Durga and Kali.

DAKSHINESHWAR (CALCUTTA)
RAMAKRISHNA

So with little thought, Joe cashed his last travelers check and took the next train north. He visited Shantiniketan and was surprised at how immaculate the tribal Santal village was. Much of rural India, and most of urban India, lacked (and still lacks) electricity, clean water, and organized sanitation. But the Santal village was truly amazing for its immaculate and organized appearance.

At the train station to Calcutta, Joe again encountered Ben Simone; the same English-speaking Jew from North Africa who had been in Pondicherry. He had also been traveling around India, meeting some of the great sages here and there.

What Joe liked most about Ben Simone was his sense of humor. He was a knowledgeable, educated, serious scholar. But when it came to India, he was also able to enjoy the ridiculous eagerness of men and women trying to satisfy their momentary desires. They literally lusted for things like television, bikes, better housing. And, like Joe, Ben Simone lived simply. He enjoyed an Indian vegetarian diet and was eager to share what he knew and what he had learned.

During the past weeks, while Joe had benefited from what he saw and did, he had missed having a like-minded soul to share his ideas and doubts. Maybe Ben Simone felt the same way, for quickly they decided to travel together to the Ramakrishna Temple in Dakshineshwar. Ben Simone explained to Joe that Sri Ramakrishna, about whom he knew a lot, was regarded much the same way Christ was accepted among the believers in the West. He had an immaculate birth and lived a life of piety and spiritual growth. At the Temple where Joe and Ben Simone headed, Ramakrishna worshipped Kali, also known as the Mother of the Universe.

Sri Ramakrishna

When they finally entered the temple, they went directly to the statue of Kali, and sat down in meditation. They sat a long time. After a while Joe got up and went outside to stretch while Ben Simone continued to sit before Kali. As Joe looked out over the River Ganges, he noticed that, whenever a train went over the bridge that spanned the river, the whole trainload of men, women, and children would all break out in a chant, "*Ma Ganga, Ma Ganga.*" The sound reverberated throughout the hills and the temple itself, producing an almost-hypnotic effect. The temple grounds were peaceful, although not well cared for, and, as Joe sat and waited for Ben Simone, someone who might have been a priest or caretaker came and asked Joe if he was interested in Sri Ramakrishna.

"Yes, certainly," Joe answered, "but my friend is more knowledgeable than I, and we are both here to learn."

"Welcome, then," the man answered. "Sri Ramakrishna welcomes those who seek to learn more about the divine, and I and a few others try to encourage visitors, but, as you can see, there are few visitors here."

Ben Simone heard the sound of the two outside and joined them with introductions all around. The man said he was Kumar but never said if he was a priest or a caretaker. It didn't matter, really, since Joe and Ben Simone were welcome to stay and offered to join him and another member of the staff at dinner that night.

"You know," Kumar said, "In India we believe that the gods take various forms, and, when they visit, we never know if they are paying us a visit in the form of a tourist or as you Joe and you Ben Simone."

"Well," Joe answered, "I can assure you--speaking for myself--that I am just learning about India and gaining a good deal of insight into my own self, but, as far as gods are concerned, I have nothing to offer."

"We'll see about that," Kumar answered. "Many people arrive here with one idea and leave quite transformed."

Ben Simone, Joe (2nd from right) and Dakshineshwar Priests

Over the next several days, visitors occasionally arrived and took tours of the temple and heard talks on the life of Sri Ramakrishna. But since Joe and Ben Simone were living there, they were asked to help with the rituals as they were performed each day. Joe enjoyed this enormously, since he had to wear a dhoti and use either a drum or bells as Kumar signaled when the occasion arose. The short talk, which often brought good contributions for the upkeep of the temple, was always well received, and, after a few days, Joe could recite much of the story himself.

Ramakrishna was born in 1836, and visions came to his parents that their son was conceived miraculously and embodied the divine even as an infant. In his youth, he was loved by everyone in his small, rural Bengal village. He could also see around him the suffering brought through fate or the sufferer's own fault. Some people were hard-hearted or selfish and interested in only their own gratification. At eighteen, Ramakrishnan entered the temple at Dakshineshwar, where his brother Ramkumar was employed. Here he took to heart the voice of the god Krishna, who speaks in the Bhagavad Gita, saying, "Whenever virtue subsides and unrighteousness prevails, I embody myself again, for the destruction of the wicked and the protection of the virtuous."

At the temple, where now they lived, Joe and Ben sat where Ramakrishna sat and asked the same questions that Ramakrishna had asked in meditation. Before the god Kali they repeated, "Dost thou really exist? Why cannot I see thee? Why dost thou not reveal thyself to me?" They both tried fasting and renounced all worldly pleasures. But for Joe it never seemed to be what he was looking for. Ben Simone explained that Joe was too programmed for enjoyment from his years growing up in New York City and, as a result, Joe would be condemned forever to seek the divine but never come to know God. It was sad to hear this from anyone, especially from a close friend with whom Joe had shared some of his sensitive thoughts. Still, Joe's nature was such that it did not end his search for redemption, and soon he would take up the search again. But now his mind turned to Japan.

JAPAN AND THE REVEREND OGATA-SAN

With his last $100, Joe bought a steamer ticket to Yokohama without a thought of what would come next. The ship stopped all along the way and may have taken two to three weeks to reach Japan. In 1949, Mao Tse Tung came to power in China and thousands of Chinese students were eager to help their new country to progress. They had beaten the Japanese, the Americans, and the hated Chiang Kai Chek puppet government. Hundreds of students and others came aboard at every stop--Rangoon, Singapore, and every place else. Now it was their turn to take over the world. And each night on the deck some student group would sing patriotic songs and put on plays about how the peasant classes under Mao's leadership had overcome the industrial might of foreigners. It was an emotional, enthusiastic, and unforgettable trip, like a festival on a world stage.

Most of the students disembarked in Hong Kong, but Joe and the other remaining seafarers proceeded directly to Yokohama. On the ship were some Japanese students who were intrigued by Joe's carefree attitude and unconcern about not knowing anyone in Japan, despite being without a dollar in his pocket. Yet, there the ship would dock in just a few days.

"But how will you live?" they wanted to know.

"Oh, I've been there before and something always turns up. We'll just have to wait and see. In the meantime, teach me a little Japanese so I can at least say hello when we dock."

There were a few easy words he picked up, *Ohayo* [hello] and *arigato gozai mashita* [thank you] and others. Finally, two students invited Joe to come with them to their university where they could give him a bed and find other students to help, and that is where Joe ended up--at Tokyo University, otherwise known as Aka Mo, or The Red Door. Joe's first introduction to Japan was the bath--a really welcome Japanese institution--and then the Sobaya shop, a place favored by the students for serving pretty much only a huge bowl of noodles: a sort of Japanese equivalent of American pizza places. To have some money, Joe borrowed $50.00 from the American Embassy and spent a number of days in a working-class bar where men drank Saki and watched Sumo wrestling on television.

RECEIPT AND PROMISE TO REPAY FUNDS ADVANCED
AS FINANCIAL ASSISTANCE LOANS FOR REPATRIATION

Tokyo, Japan
May 23, 1955

I, Joseph Di BONA, have today received from the American
Embassy at Tokyo, Japan, the sum of $50.00 equivalent to ¥18,000
which I promise to repay without interest to the Treasurer of the
United States upon demand, in legal tender of the United States.

That sum is required for my subsistence from the period from
May 23, 1955 to May 30, 1955.

I understand that my obligation to repay the sum herein stated
will not be discharged until the Treasurer of the United States
actually receives in legal tender of the United States full repayment
of that sum.

Repayment of the amount of this loan may be obtained by the
Department of State from any monies now or hereafter due me from
the Veterans Administration or other agency of the United States
Government.

I further understand and agree that after my repatriation
I will not be furnished a passport for travel abroad until my
obligation to reimburse the Treasurer of the United States is
liquidated.

SIGNATURE OF APPLICANT: _____
/ Joseph Di Bona

LOCAL ADDRESS: c/o American Express, Tokyo, Japan.

ADDRESS IN THE UNITED STATES: c/o Mrs. Antoinette Ranatti
17 Terry Street, Hicksville
Long Island, New York.

_____
(Witness)
Henry E. Mulloy

113

With the government loan, Joe found his way to Kyoto, where he had an introduction from Sivananda to a Reverend Ogata. He was a Zen priest who presided over Shokokuji Temple in Kyoto. It was a small but exquisitely beautiful temple. Without a word more, Joe was invited to stay in Ogata's home and share his meals. It was the most gorgeous place imaginable, or at least as beautiful as Joe would ever know. His room had only a tatami floor with sliding paper walls that opened onto a scene of flowers and trees above which rose the snow-clad mountains in the distance. The only thing in the room was a lovely vase with a single iris. After half a century, the simple beauty still takes Joe's breath away. In that lovely room, he would take out his spinning wheel and spin while looking out on that amazing scene. One of Ogata-san's daughters wanted to learn spinning, and Joe was able to teach her very quickly.

In the daytime after breakfast, Joe wandered around the city, amazed at the cleanliness and good taste everywhere. Even in the poorest parts of Kyoto, he would see lovely flowers carefully arranged to capture his eye or his heart. One day he walked down the length of Karasuma Dori Road from the Shokoku-ji Temple to the Kinkoku-tei Garden, a lonely spot without paths. The grass had over grown the area for want of visitors. Joe photographed the antique wooden bridge beside the pond and entered the Marubutsu Department

Store. There he saw a large dining room with dummy models of the dishes being served. Thanks to his Chinese, he was able to read the prices. Things went well. He met a French guide, who had no one who needed his services, so he offered to take Joe around, just to practice his French. They stopped at the Minamiya Theatre, where they found a performance of The Romance of Genji. It could have been anything, for all Joe understood of the action, but the costumes and movements were gorgeous, and the faces and music exquisite. Kyoto is a large city with lots to see, but Joe would have preferred a smaller city with more time to see the sites.

Room in Kyoto and spinning wheel

On another day, some children showed up at Joe's room to watch him spinning, or maybe it was his beard that intrigued them. After they left, Reverend Ogata told Joe that he wanted to take Zen to America, and Joe responded that the one point of contact for all religions was meditation. There everyone begins equal and encounters the same problem: "What am I expected to find? How shall I concentrate? On what? If it is non-intellectual, non-rational, nonconceptual, how can I even conceive it? Can I really achieve unity with the infinite?"

The next morning Joe took a tram from Shokoku-ji and as far as the Kyoto Museum, where he walked away, passing some houses of prostitution, where the women wore white and called to the passers-by. He wanted to go in, curious to see what it was like, but decided against it. Money was what it was all about, and if the man did not buy, the women would be unhappy. The houses were completely business establishments, and not at all like Palermo--where a man can just enter, sit down, have a smoke, see the girls, talk to them, and leave without having spent any money at all.

On yet another morning--April 8, 1955--Joe woke late, and, when he went to wash, Ogata-san spoke to him. He had obviously been waiting for Joe to get up, for he had something important to say. This was the day of the investiture ceremony for the new abbot of Myoshin-ji Temple, and Ogata had been invited to attend. Joe could accompany him if he wished. It was a casual invitation, but one no one would easily forgo. Myoshin-ji is the largest Rinzai Zen center in Japan and the abbot has absolute control over some 5000 souls all over Japan. Formerly the investiture ceremony was of such importance that an Imperial envoy was sent to attend. However, since meiji times, this practice has been discontinued.

Ogata-san at the Investiture

Ogata was already dressing in his official robes: white under kimono, then rich blue on which he wore a sort of gold brocade apron, and, over all, a darker Kimono for street wear. On his feet were white socks and wooden shoes, and, in his hand, only his ceremonial fan. When they arrived, the procession was just beginning, with a visit to the founder's tomb. In front and behind the new abbot were lines of monks, dressed much like Ogata, all with shaved heads and perhaps an extra silk toga or two. The new abbot, himself in gold and under an orange palanquin, was surrounded by little five and six year olds, made up much like one might imagine the mikado himself might dress. Their faces were powdered white, with reddened lips and eyes painted on the little pagodas of tinseled gold on their heads.

The monks lined up in the Founders Hall, a building said to date from about 1400 years ago when Buddhism first came to Japan. The building is supposed to be Tang, or at least Sung, architecture and, in fact, the whole ceremony is strongly Chinese. For example, the head monks wear black slippers embroidered with white, said to be exact replicas of Sung footwear. In the temple grounds is a bell brought to Japan by the first Chinese Buddhists and renowned for the primary tone of the Chinese musical scale. In the ritual, everything proceeds very slowly and deliberately. When the abbot chants, each word is prolonged for a full second or two, at least. After the visit to the Founder's Hall, everyone assembles in the main building. Joe found that most of the Japanese visitors were formally dressed in morning clothes, the women elegant in their best obis and kimonos.

Rinzai Zen Abbot

Joe took a place at the back so as to be able to stand up from time to time, whenever his legs might became tired of squatting. Always, slowly, slowly, the ceremony proceeded. A sort of Major Domo mounted the dais to give the new Abbot something. He took a step and stopped, took a step and stopped. Up, down, and the minutes passed. The pace was new for Joe and had a charm all its own. If not charm, it had a hypnotic effect on everyone. Joe did not feel it should go faster. He felt that would have been impossible, and, little by little, once he accepted that fact, the entire affair became quite natural. Then he and the other onlookers entered into the measured paces and prolonged vowels. It was another world realized, so perfectly presented that the participants and visitors were indeed forced to relive the same ceremony of the past 1400 years.

Later, back in his room at the temple, Joe had another visitor: a young boy intrigued with the spinning. Joe taught him a little on the takli, and then gave him one and some cotton to practice on. Just as the boy was leaving, a panel slipped open and a little girl announced, "Bath is ready."

"What?" Joe asked.

"Bath is ready." And so he bathed.

After a few days spent preparing to leave Japan for San Francisco, Joe had to make his good-byes to all those whose generous kindness he had enjoyed. On the evening before he was to leave, he asked Ogata-san for instruction on meditation. Ogata-san agreed. They had eaten and bathed and were rested. They met in the main meditation hall, which had a raised platform in front. It was dusk outside and only a candle illuminated the area, giving it a certain mysterious quality. When Joe was sitting on the platform, Ogata came in and asked him to cross his legs as Ogata crossed his. Joe did so without difficulty. Then Ogata asked him to hold his hands in a particular position, and he did. Then he was told to concentrate his mind on the front of his temple. Joe tried to concentrate, but he was so distracted by the novelty of what they were up to and his upcoming travels that he could not.

They sat there for a while, their eyes closed, trying to concentrate. Then Ogata spoke, saying, "I have been doing this for many years and nothing ever happens." Joe was shocked to hear the great Zen master confess that it was all a hoax. Or was it? He could not fathom what Ogata really meant, but then he remembered the long Zen tradition of using such odd ideas to instruct the novice. Didn't Zen masters laugh at funerals, and what were koans all about, anyway? What is the sound of one hand clapping? You tell me! And here was another example of sidestepping the usual heavy bleakness of religious instruction. Was that it? Was Ogata telling Joe that life is merely a joke, with life and death all the same? Joe left perplexed but more accepting of how to see the world. The next day, Joe took his leave of Japan.

One of the conditions the U.S. embassy asked for when they lent Joe the money was that he write to all his relatives and friends, asking for their help in getting back to the States. Not only did he not hear from anyone who might provide funds, but he even received some very unsympathetic mail, telling him why he did not deserve assistance. The gist of the letters went something like this: "You got yourself into this hole, and you should figure a way to get yourself out. Besides, you have pretty much been having a ball; while we back here at home in the good old U.S. of A. have been working our bottoms off, trying to keep afloat. We know from your letters that people have helped you all along, and we don't see how you have ever suffered in any way. And spinning cotton, what is that supposed to be? Your sister has been a nurse all these years, her husband Bob is at Grumman on Long Island, Tom is a photographer, and Ann is a seamstress, and you? What do you do?--You spin cotton, and now you want us to bail you out! Think again, Joey."

This notion represents a composite attitude. They seemed to resent Joe's idea of retiring in his 20s and enjoying traveling around the world and only later working until he was in his 80s. That was such a strange idea that they could not begin to understand it.

However, there was one exception and that is how Joe got back to America. A woman Joe had gone to the University of Wisconsin with, and knew again in Europe, sent him $500 for a boat ride to San Francisco. He was not entirely surprised, nor did he resent the naysayers. In a way, he had optimistically expected that everything would work out. He was living a charmed life, and had been for a long time, so what was there to worry about?

When his ship arrived in San Francisco, there was this one woman on the dock to greet him. It was raining, and she wore a tan raincoat and hat. As he saw her there, he was almost moved to tears. "This is almost miraculous," he thought. As soon as he was off the gangplank, he embraced her, kissed her ardently, and asked her to marry him. She accepted on the spot and thus this part of Joe's life ends and a new chapter begins.

Helene, summer 1955

EPILOGUE

Most epilogues are not worth reading. They are long, boring, unnecessary, and can be omitted without damage to the text you have already read. This one may be different in so far as it brings together the vast differences that were perceived between the rural culture of India and the excitement of Berkeley, California.

Yes, I am the "Joe" of the narrative, and, after we married, Helene and I moved into a rented house near the university so Helene could walk to work. She taught English and was our sole source of money to live on. I was not working nor did I intend to compromise my principles, which included self-sufficiency and finding personal growth through spinning and following Ghandian values. Nonetheless, since it was difficult to live on Helene's meager salary alone, I realized that I had to earn some kind of income. I refused to

Helene and Joe's wedding, July 1955

work for any company that had employees because I regarded capitalism as exploitation and, therefore, evil. But I liked both marriage and California, so I looked around for something to do.

I found it in a large house next to the university, which rented at the time for $125.00 per month. It had four rooms upstairs, so I fixed them up and rented them to students, each for $30.00 per month. Helene and I lived in the lower half, and, very quickly, we also had to make room for our first child--a lovely girl whom we named Nicole. Two years later we added a son, Pete, to our number. Fatherhood made me quickly realize that, with Helene's salary unlikely to increase with any speed, I really had to earn more for my family.

But, first, I must say something about my attitude toward food, eating, and the restaurant industry around us. I had been influenced—no, a more accurate phrase is totally consumed-- by what I experienced in Sevagram. For instance, the ashram had had no refrigerators. In fact, there had been no electricity at all, and no one seemed to be concerned about that fact. We had all lived in a perfectly happy village society and everything had gone completely well for us in Sevagram. Our

124

meals there were prepared daily, fresh and tasty and nutritious with usually nothing left over. So, when I arrived in California, I asked myself, "Why do we need a frig at all? I will shop daily for our food and we will prepare it as best we can."

Helene was brave and willing to indulge my somewhat unorthodox ideas. At that point, she had had no experience cooking, nor had her mother. In her home it was her grandmother who was in charge of the kitchen, and jealously guarded her right to control whatever was cooked and whatever was served. As a result, neither my wife Helene, nor her mother, knew anything about food. I must say that Grandma Moses was from a French Jewish family and had mastered the most wonderful sauces and types of food preparation imaginable. So whenever we visited Helene's parents and her live-in grandmother, dining was always a memorable experience.

Still, it was not long before the pressures of American society took their toll, and I succumbed to finding a place for infant formula and other needs I was growing used to. (Do I need to go into a developing passion for cold beer?) But how were we to pay for all of these modern American "conveniences"? I looked around and found a gardener's position at the art college in Oakland and I thought, "This is great!" The college soon offered me an additional job of repairing books in the college library. And this offer I quickly accepted. I worked under the guidance of an elderly expert who made her own glue, used string to tie new bindings together, and did all the other things necessary to repair books. I loved the work and believed that if I ever mastered the skill, bookbinding would turn out to be my true calling in America.

So I worked at the two jobs, gardening and book repair. Luck was still with me, for the college officials decided, rather than spend money to increase my salary, to give me, instead, a tuition-free grant towards an MFA. Again I accepted their offer, and quickly developed a new passion for making ceramics, which became my major.

I grew so fascinated with clay, glazes, and the way the fires of a kiln seemed to purge away earthly desires that I would stay up all night, watching the fire melt the glazes. Eventually, I needed to work on my MFA thesis, so, still remained focused on simplicity, I researched the Pueblo American Indian potters of the American Southwest. The Pueblo used cow dung as fuel, employed clays native to where they lived, and created art forms closely related to their actual lives.

My completion of my thesis on these potters and of a few creative pieces of my own brought me my Master of Fine Arts degree from the California College of Arts and Crafts. With that degree, I was able to teach ceramics for the recreation departments in the Berkeley area. In my role as ceramics teacher, I fancied myself weaning people away from the institutionalized and alienating forms that had no relation to their lives. I asked them to think of what they needed rather than succumb to advertised ideas that only enriched corporate America.

But while all this was going on, I became increasingly interested in politics and economics. Here, too, I was influenced by my Indian sojourn, by Gandhi and Vinoba Bhave in particular. While I was in Sevagram, there was a heavy argument about spinning. One faction pushed for self-sufficiency; the other called for increased production and its consequent additional revenue. I have earlier noted that the ashram has both good and mediocre spinners. The mediocre spinners simply could not spin well enough to bring needed income into the ashram.

To handle this problem, the amber charka was invented. This device was still controlled by a single spinner, but instead of spinning only one thread the spinner could produce two or more threads at the same time and, thus, multiply output. The argument expanded to include moving the group ahead in all industrial production, versus returning society to its earlier, more primitive state. In Sevagram and, really, among most of India's intelligentsia, this debate between accepting modernization versus maintaining primitive self-sufficiency raged on, but in Berkeley, California, the notion of returning to the simple, primitive ways of living seemed unnatural and, really, only appealed to very few. Still, I was so enamored by the joys and importance of self-sufficiency that I determined to help expand the wisdom of Sevagram, but now in California.

I ordered a dozen spinning wheels from India and, when they arrived, I began recruiting people to learn spinning. I found that one of my fellow college students, Kay Sekimachi, was interested. And she recruited her sister, Koz, and then their mother. The two girls were accomplished weavers and one became nationally famous in her craft. Soon we were about five or six strong, all learning to spin. We went as far as to find an empty store, where we sat in such a way that people could pass and watch us spinning. Our spinning always fascinated the public, and over time more people joined us. A problem arouse in getting a supply of cotton. Loads of cotton is grown in California and, after it is mechanically harvested, a great deal of cotton is still left in the field. Our group would go out and glean more than we could handle. Back home, we would card the cotton, which is taking the seeds out, and then fluff and roll the batters for spinning.

It is a long and slow process, but our group learned to enjoy the new pace that affected other aspects of the life. We never did make enough yarn for a piece of clothing, nor did we manage to end the domination of large industry in America.

But I believe that our own lives were immeasurably enriched by the experience.

While this cotton spinning was going on, another activity began taking more of my time: commenting on government, politics, and international affairs. Toward that end, I constructed a traveling theatre, ostensibly as an entertainment form. Part of the physical "theatre" was a frame-and-curtain suspended from a rigid roof that I attached around my head. Two puppets appeared in front of a curtain and "discussed" public affairs with people. Both young and old alike would argue against or endorse the issues the puppets raised, and everyone would have a good time. The only time I took out the mobile theatre was during a fair or a parade or when a crowd was present, but the puppetry caught on and always created good publicity for our causes.

After graduating from the art college, I decided to teach in the public schools and studied in the Berkeley School of Education. I was not happy with the curriculum. For example, in one class, which seemed to demand some musical ability in drumming or ringing bells, I was given a

Gleaning cotton in California

low grade. In another class, we had to fill out a pseudo class roster and note who was absent and other facts for the use of school authorities. In this exercise, too, I proved less proficient and was in danger of failing when I was unexpectedly and happily hired before the course ended and automatically endorsed by the school where I planned to teach, despite my poor grades. My first year of teaching was exciting, but my second year seemed to be pretty much a repeat of what I had done the previous year. The third year was fairly clearly a mistake.

Next, I applied and received a grant to study South Asian Studies at the Berkeley University Graduate School. My Berkeley education led to more grants for research in India. I returned to India in about 1963 with money to research the Indian university. I settled on Allahabad, an important educational center in north India. When I arrived at my chosen Indian university, I found the students on strike, not attending class and threatening anyone who defied their efforts to bring the university to a total halt. I quickly changed my research suddenly to studying student politics and not anything else. At that moment there was a great interest in students and regime changes around the world, as well as at Berkeley, and, as a result of my research topic, I had no trouble finding an academic job.

Allahabad, where I spent over a year, was a quiet provincial town dominated by its major university, religious importance, and the provincial high court. What was unusual was the way my small monthly stipend enabled me to live like a prince. I could afford a fine house, a cook, a gardener, and anything else I needed or wanted. For the first time in my life, I had more money than I could possibly spend. With such wealth, I brought my wife and two small children to join me in Allahabad. Unfortunately my wife, Helene, became seriously ill and had to return to America.

Joe, Nicole and Pete in Hawaii

Suddenly, I was left with my research to complete and with looking after Nicole, who was eight, and Pete, who was almost six. They were enrolled in private schools, and, thanks to helpful neighbors, we managed reasonably well.

When we said goodbye to India and returned to the University of Hawaii, where I still had six months to serve on my grant, I was shocked to discover that the $250 that we had lived on so lavishly in India would not begin to support us in Honolulu. Hawaii turned out to be the most expensive place to live on earth. I was devastated, not for me, but for the children, who could not understand what was happening.

The only place I could afford to rent was full of drug addicts, alcoholics, and criminals. What to

131

do? Again I enrolled the children in a parochial school and tried washing their clothes in the kitchen sink. And with a borrowed iron, I did what I could with their school uniforms. Quickly, some other families recognized our plight and came to our aid. I recall particularly a native Hawaiian family who daily gave the kids a snack when they returned from school, taught them Hawaiian songs, and more. My daughter learned to string flowers to make leis, and both children could do a dance using simple gestures. With the family's help, thank God, we survived. Perhaps I should even say thrived.

Things improved and I finished my Ph.D. in Comparative Education and South Asian Studies, and eventually got academic jobs—first at Brooklyn College and then at Duke University. Of course, there is more to say and, perhaps, at another time I will write it. But for the moment, this is enough.

List of Illustrations

List of Maps

Paris to Bombay

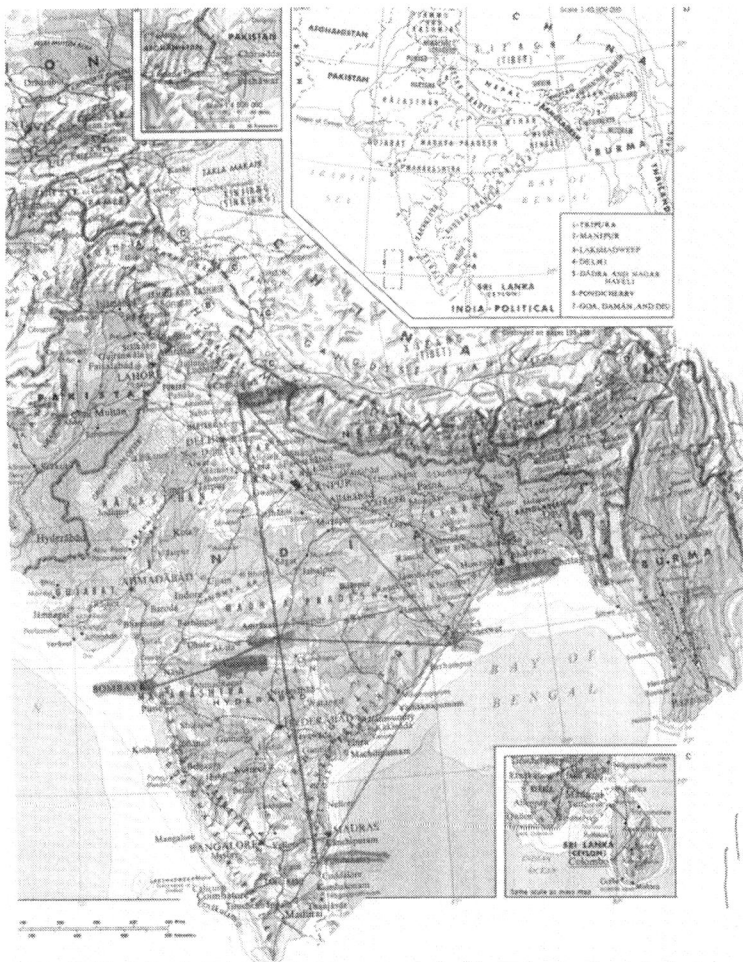

Places visited in India

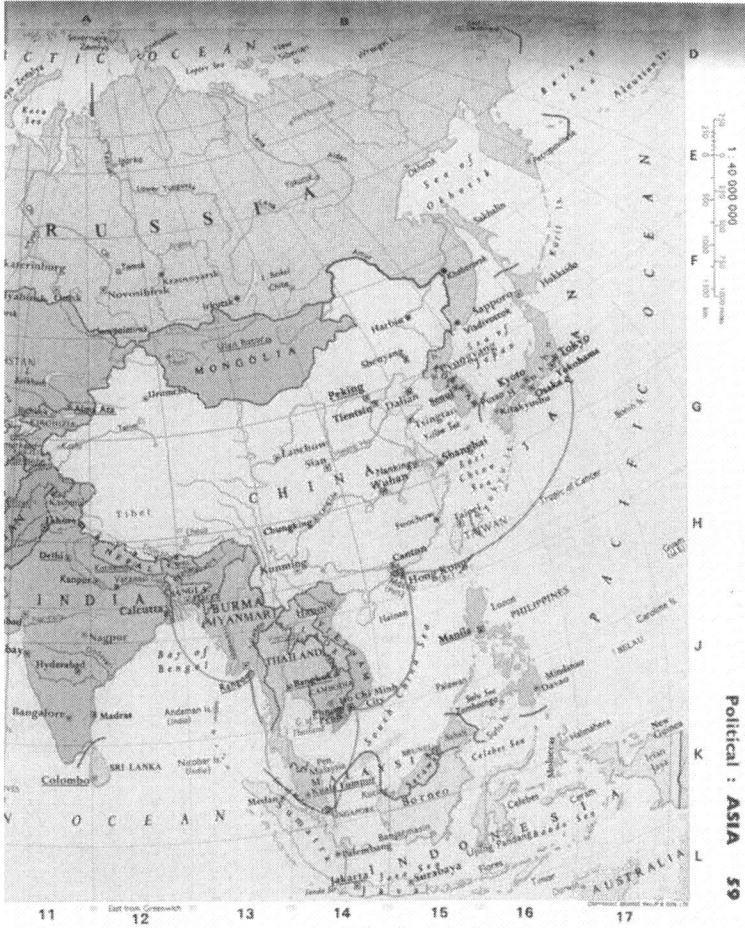

Calcutta to Tokyo

Other Books Published by the Author:

1969. *Change and Conflict in the Indian University*. Durham, NC: Duke University Program in Comparative Studies on Southern Asia. (Dedicated "To Nickle Pickle and Pizza Pie.")

1970. *Language Change and Modernization*. Ahmeddabad: Gujurat Vidyapith.

1974. *The Context of Education in Indian Development*. Edited. Durham, NC: Duke University Program in Comparative Studies on Southern Asia. (Dedicated "To Canu and Claudia. People of Two Worlds.")

1977. *A is for Ambrosia: An Alphabet Book for Smart Kids*. (Dedicated "For Vito, Canu, Pietro, and Nicole")

1983. *One Teacher, One School: The Adam Reports on Indigenous Education in Nineteenth-Century India*. Edited and Introduced. New

Delhi: Biblia Impex Private, Ltd. (Dedicated "To my staunch pals, Canu Cassio and Vito Lorenzo")

1989. *Critical Perspectives on Indian Education.* New Delhi: Bahri Publications.

2005. *My Heart Belongs to Sammy: The Kid Who Defies Convention.* Chapel Hill: Professional Press. (Dedicated "To Stacey and Alfonzo who I hope will grow up brave and resourceful, competent and happy.")

2008. *A Wayward Journey of Love and Dreams.* Righter Publishing Company.

About the Cover Artist

The cover of this book is a watercolor by Gordon Christopher, an internationally recognized artist with numerous one-man shows and whose work is represented in collections in Canada, the United States, and the Caribbean. Born in Antigua, West Indies, Gordon studied art at Alberta College of Art and at the University of Calgary. The cover is from a new series of paintings celebrating the excitement of Carnival in Antigua and is labeled "On St. John's Street." Gordon continues to explore new dimensions of his art in both Calgary, Canada, and in Roxboro, North Carolina.

About the Author

Joe Di Bona was born at home in the family's First Avenue Manhattan tenement, not far from the East River. Growing up in Yorkville, he knew the street gangs that ran the neighborhood as well as the cultural excitement of the Museum of Natural History and the Metropolitan Museum of Art. He attended P.S. 82 through the 8th grade and, in 1945, received a diploma from Stuyvesant High School. Shortly thereafter, he was in the Army and in Korea, disarming the Japanese. There was a great deal of dislocation and suffering among the Korean population, and Di Bona came face to face with hunger and poverty for the first time.

At the University of Wisconsin, he studied Political Science and then worked two unremarkable years for the Ford Motor Company in New Jersey. Seeking something different, he entered the *Ecole des Langues Orientales Vivantes,* in Paris to learn Chinese, but, instead, became sidetracked by a number of competing interests. One of the most memorable was a visit with George Roux, a holy figure in the South of France, whose followers recognized him as Jesus returned to earth. At the Roux center, Di Bona was baptized with the gift of healing by laying on of hands and always remains grateful for that gift.

In the 1950s, he wandered around the world, thinking it better to be footloose and fancy free in our 20s rather than when we are unable to enjoy ourselves after age 80. The people he met in those travels remain significant influences in his life. Anton Biro, for example, a Hungarian artist and mystic living in Paris, advised Di Bona to go to India. Di Bona asked him what to do about his lack of money, and Biro said, "Since everyone is poor in India, go there and you will feel right at home."

Righter Publishing Company

Fresh, delightful stories for the modern
reader.

Thrillers, detective stories, short story
collections, children's books, inspirational
works, poetry collections, science fiction,
romance, local histories, personal memoirs,
self-help and unintentionally educational.

Go to www.righterbooks.com

Made in the USA
Charleston, SC
15 August 2010